Springer Series in Materials Science

Volume 201

For further volumes:
http://www.springer.com/series/856

The Springer Series in Materials Science covers the complete spectrum of materials physics, including fundamental principles, physical properties, materials theory and design. Recognizing the increasing importance of materials science in future device technologies, the book titles in this series reflect the state-of-the-art in understanding and controlling the structure and properties of all important classes of materials.

Mirza Bichurin · Vladimir Petrov

Modeling of Magnetoelectric Effects in Composites

Mirza Bichurin
Vladimir Petrov
Novgorod State University
Veliky Novgorod
Russia

ISSN 0933-033X ISSN 2196-2812 (electronic)
ISBN 978-94-024-0347-3 ISBN 978-94-017-9156-4 (eBook)
DOI 10.1007/978-94-017-9156-4
Springer Dordrecht Heidelberg New York London

Softcover reprint of the hardcover 1st edition 2014

Printed on acid-free paper

Springer is part of Springer Science+Business Media (www.springer.com)

Preface

Magnetoelectric composites, which simultaneously are ferroelectric and ferromagnetic, have recently stimulated a sharply increasing number of research activities for their scientific interest and significant technological promise in the novel multifunctional devices. Magnetoelectric responses of single-phase compounds are relatively weak and occur at temperatures too low for practical use. On the contrary, composites typically show giant magnetoelectric coupling response above room temperature and are ready for technological applications. In such composites, the magnetoelectric effect is a product property of a magnetostrictive and a piezoelectric substance. Achievement of high magnetoelectric voltage coefficients necessary for engineering applications has been enabled by the appropriate choice of phases with high magnetostriction and piezoelectricity. The authors of this book have attempted to bring together numerous contributions to modeling of ME composites. They present to readers new to the field modern approaches of the physics of composites, so that the potential of the field can be made transparent to the new generations of talent to advance the subject matter.

Interestingly, the coupling interaction between nanosized ferroelectric and magnetic oxides is also responsible for the magnetoelectric effect in nano-structures, as was the case in bulk composites. The availability of high-quality nano-structured composites makes it easier to tailor their properties through epitaxial strain and interfacial coupling. In this book, the authors discuss these magneto-electric composites from both experimental and theoretical perspectives.

"Modeling of Magnetoelectric Effects in Composites" is an excellent text covering fundamentals of analytical modeling, material behavior, and experiments. The leading authors in this field provide an in-depth coverage of modeling principles for refining the performance of two-phase architectures over a wide frequency range. The book is a must read for researchers investigating new connectivity patterns in piezoelectric-magnetostrictive materials.

Yu Gulyaev

Contents

Abstract

This work is dedicated to the theoretical modeling of magnetoelectric (ME) effects in layered and bulk composites based on magnetostrictive and piezoelectric materials. Our analysis rests on the simultaneous solution of elastodynamic or elastostatic and electro/magnetostatic equations. The expressions for ME coefficients as the functions of material parameters and volume fractions of components are obtained. Longitudinal, transverse, and in-plane cases are considered. The use of the offered model has allowed to present the ME effect in ferrite cobalt–barium titanate, ferrite cobalt–PZT, ferrite nickel–PZT, lanthanum strontium manganite–PZT composites adequately. It should be noted that as opposed to well-known works on this topic, in this study we showed the way of calculation of ME parameters in detail and developed all examples to numerical estimations.

Since ME coupling in the composites is mediated by mechanical stress, one would expect orders of magnitude stronger coupling when the frequency of the ac field is tuned to acoustic mode frequencies in the sample than at nonresonance frequencies. A model is presented for the increase in ME coupling in magnetostrictive-piezoelectric bilayers for the longitudinal, radial, bending, and shear modes in the electromechanical resonance region. We solved the equation of medium motion taking into account the magnetostatic and elastostatic equations, constitutive equations, Hooke's law, and boundary conditions. To obtain the inverse ME effect, a pick-up coil wound around the sample is used to measure the ME voltage due to the change in the magnetic induction in magnetostrictive phase. The measured static magnetic field dependence of ME voltage has been attributed to the variation in the piezomagnetic coefficient for magnetic layer. The frequency dependence of the ME voltage shows a resonance character due to acoustic modes in the sample. The model is applied to specific cases of cobalt ferrite–PZT, nickel/permendur–PZT, and metglas–PZT bilayers. Theoretical ME voltage coefficients versus frequency profiles are in agreement with data. In addition, we consider a composite based on magnetization-graded magnetic phase from compositionally graded ferromagnetic and polarization-graded piezoelectric phase from compositionally graded ferroelectric.

We provide the theory for ME effects at the coincidence of ferromagnetic and electromechanical resonance, at magnetoacoustic resonance. At ferromagnetic resonance, spin-lattice coupling and spin waves that couple energy to phonons through relaxation processes are expected to enhance the piezoelectric and ME interactions. Further strengthening of ME coupling is expected at the overlap of ferromagnetic and electromechanical resonance. We consider bilayers with low-loss ferrites such as nickel ferrite or YIG that would facilitate observation of the effects predicted in this work. For calculation we use equations of motion for the piezoelectric and magnetostrictive phases and equations of motion for the magnetization. The ME effect at magnetoacoustic resonance can be utilized for the realization of miniature/nanosensors and transducers operating at high frequencies since the coincidence is predicted to occur at microwave frequencies in the bilayers.

Chapter 1
Magnetoelectric Interaction in Solids

Abstract Magnetoelectric (ME) interaction in magnetically ordered materials is reviewed. To create new magnetoelectric composites with enhanced ME couplings we discuss the ME properties of ferrite-piezoelectric composites. Such materials would enable one to make novel functional electronics devices. The main objective is a comparative analysis of ME composites that have different connectivity types. It is to emphasize that multilayer composites possess giant ME responses and at the same time the relative simplicity of manufacturing. In addition, composites with 3–0 and 0–3 connectivity types are also easy to make using a minimum monitoring of the synthesis process. The ultimate purpose of theoretical estimates is to predict the ME susceptibility and ME voltage coefficients as the most basic parameters of magnetoelectricity. The magnetoelectric effects occur over a broad frequency bandwidth, extending from the static to millimeter ranges. This offers important opportunities in potential device applications.

The magnetoelectric (ME) effect in a material consists in inducing an electric polarization by an applied external magnetic field, or vice versa in inducing a magnetization by external electric field. Linear state functions that define these cross couplings can be given respectively as

$$P_i = \alpha_{ij} H_j, \tag{1.1}$$

$$M_i = \alpha_{ji} / \mu_0 E_j, \tag{1.2}$$

where P_i is the electric polarization, M_i the magnetization, E_j and H_j the electric and magnetic fields, α_{ij} the ME susceptibility tensor, and μ_0 the permeability of vacuum. The ME effect in solids was first theoretically predicted by Landau and Lifshitz (1980), calculated by Dzyaloshinskii in Cr_2O_3 (Dzyaloshinskii 1960), and experimentally observed by Astrov (1961) and Folen et al. (1961).

When uniform magnetic and/or electric fields are applied to a material, the change in the Gibbs free energy density can be expressed as (Landau and Lifshitz 1980)

M. Bichurin and V. Petrov, *Modeling of Magnetoelectric Effects in Composites*, Springer Series in Materials Science 201, DOI: 10.1007/978-94-017-9156-4_1,

$$dF = -P_i dE_i - \mu_0 M_i dH_i \tag{1.3}$$

Using 1.3 enables one to obtain the following thermodynamic relationships for the dielectric polarization and magnetization:

$$P_i = -(\partial F/\partial E_i)_{H,T}, \tag{1.4}$$

$$\mu_0 M_i = -(\partial F/\partial H_i)_{E,T}; \tag{1.5}$$

where T is the absolute temperature. If we assume that the electric susceptibility (χ^{E}) and magnetic susceptibility (χ^{M}) are, respectively, independent of their primary ordering fields E and H, we can obtain the following free energy expression for a linear dielectric and magnetic system that has an ME exchange between the subsystems, given as

$$F = -1/2\,\chi_{ij}^{E} E_i E_j - 1/2\,\chi_{ij}^{M} H_i H_j - \alpha_{ij} E_i H_j. \tag{1.6}$$

The first term on the right is the electrical energy stored on application of E, the second the magnetic energy stored on application of H, and the third the bilinear coupling between the magnetic and polar subsystems on simultaneous application of E and H. From 1.6, we can then obtain the following expressions for the total polarization and magnetization induced by simultaneous application of E and H:

$$P_i = \chi_{ij}^{E} E_j + \alpha_{ij} H_j, \tag{1.7}$$

$$M_i = \chi_{ij}^{M} H_j + \alpha_{ji}/\mu_0 E_j. \tag{1.8}$$

The ME susceptibility is a second-rank axial tensor, involving an exchange between polar (E_j) and axial (H_j) vectors. It is unique in these regards to the dielectric susceptibility and magnetic permeability, which are both second rank polar tensors. This is an important point, because as a consequence, the values of the components of the ME susceptibility tensor will be dependent on the magnetic point group symmetry, rather than merely on the crystallographic one.

Dzyaloshinskii theoretically showed that among materials of known magnetic point group symmetry that there was at least one crystal—chrome oxide—in which a ME effect (Dzyaloshinskii 1960) should be observed. In 1960, Astrov experimentally verified this previously predicted ME coupling in Cr_2O_3 (Astrov 1961), and reported the values of the longitudinal and transverse ME susceptibility coefficients that are given by 1.2. These measurements were performed by measuring the ac magnetic moment induced in the sample by application of an ac electric field driven at a frequency of $f = 10$ kHz. Subsequently, Folen et al. (1961) reported the magnetic field induced polarization of Cr_2O_3 given by 1.1. Said studies were done using a simple procedure. Electrodes were deposited on both ends of a single crystal bar-shaped sample. The crystals were then located between the poles of an electromagnet, placed in a vacuum chamber, and an electrometer was used to collect the charge induced across the crystal in response to an applied magnetic field. Shortly, thereafter, this same measurement methodology was used by Asher (1969) to study $Ni_3B_7O_{13}I$.

The structure–property relations of the matrix of ME coefficients has been studied for some systems (Folen et al. 1961). Classification of ME materials has been carried out for various magnetic point groups (Asher 1969), and the symmetry of ME properties determined by application of Neumann's law (Santoro and Newnham 1966). It is known that the property matrix is null except for the case of magnetically ordered materials: in which case, the matrix of coefficients is antisymmetric. Theoretical models based on experimental ME data have been presented for a few material systems, and interpretation of the ME effect have been attempted using statistical and phenomenological approaches. A Landau-Ginzburg approach has been also applied to study ME effects in ferrimagnets (Aubert 1982), where the magnetic energy density is decomposed into a power series in terms of the spontaneous magnetization and the spatial derivative thereof. In this case, there is a specific symmetry relationship between the matrix of coefficients in the higher (paramagnetic) and lower (ferromagnetic) temperature phases.

Electromagnetic wave transmission in a ME media has been studied (Opechovski 1975; O'Dell 1970). Brown et al. (1968) have reported a theoretical value for the upper limit of the ME susceptibility. These authors showed that

$$F + (1/2)\chi_{jj}^{d} H_i^2 \leq 0; \tag{1.9}$$

where χ_{jj}^{d} is the diamagnetic susceptibility. By taking into consideration of 1.6, the following inequality results:

$$(1/2)\chi_{ii}^{E} E_i^2 + \alpha_{ij} E_i H_j + (1/2)\chi_{jj}^{p} H_j^2 \geq 0; \tag{1.10}$$

where $\chi_{jj}^{p} = \chi_{jj}^{m} - \chi_{jj}^{d}$ is the paramagnetic susceptibility. Taking into account that 1.10 is a positive-defined inequality, and that $\chi_{ii}^{\mathrm{E}} \geq 0$ and $\chi_{jj}^{\mathrm{p}} \geq 0$, one can arrive at

$$\alpha_{ij} < \left(\chi_{ii}^{E} \chi_{jj}^{p}\right)^{1/2}. \tag{1.11}$$

If we now suppose that the diamagnetic component is small relative to the paramagnetic one (i.e., we limit ourselves to the case of materials with localized magnet moments), we must consider the case that $\chi_{jj}^{\mathrm{p}} \approx \chi_{jj}^{\mathrm{m}}$, and therefore

$$\alpha_{ij} < \left(\chi_{ii}^{E} \chi_{jj}^{m}\right)^{1/2}. \tag{1.12}$$

An analogous relation can be obtained on the basis of thermodynamic theory (Brown et al. 1968)

$$\alpha_{ij} < \left(\varepsilon_{ii}\mu_{jj}\right)^{1/2}; \quad (1.13)$$

where ε and μ are relative dielectric permittivity and magnetic permeability, respectively. For known material systems, the theoretical upper limit is known to exceed the experimentally measured values.

In addition, a linear ME effect was discovered in ferrimagnet $Ga_{2-x}Fe_xO_3$ (Freeman and Schmid 1975). Interestingly, in this unusual system, this ME effect was explained by the simultaneous presence of piezoelectric and piezomagnetic properties, where one striction effect was coupled to the other. In Ferrum borate $FeBO_3$ (Bichurin and Petrov 1987), quadratic ME effects mediated by piezoelectric–piezomagnetic dual couplings has also been reported. These were the first results which suggested that the Dzyaloshinsky-Moriya spin-lattice exchange may not be the only mechanism by which ME effects could be achieved: it was an intellectual precursor to magneto-elasto-electric interactions in composites.

From the 1950s to the present time, many single-phase ME single crystals have been studied (Smolenskii and Chupis 1982; Schmid et al. 1993; Bichurin 1997; Bichurin 2002; Fiebig et al. 2004; Rivera 2009; Wang et al. 2009). Most of these materials have ME effects only at temperature considerably below that of 300 K. This is due to materials having either a low Neel temperature or a low Curie point: there have been no reports of a material with simultaneously large polarizations and magnetizations at or above room temperature. This is important because the ME tensor coefficients are vanquished as soon as the temperature approaches the transition point at which one order parameter transforms to the disordered (para) phase. Furthermore, even at low temperatures, single-phase ME crystals have very small values of the ME coefficients: making them impotent in practical use. Fortunately, composite materials of ferrite and piezoelectric phases exist which have simultaneously large polarizations and magnetizations to temperatures much higher than normal ambient ones: offering to date the only practical approach to ME applications. We now turn our attention in the remainder of this book to ME composites, which are based on magneto-elasto-electric interactions.

1.1 Magnetoelectric Coupling in Composites

Composite materials offer the opportunity to engineer properties that are not available within any of its constituent phases. First, composites have the conventional group of colligative properties which includes, for example, density and stiffness: where composite quantitative adjectives are determined by individual component adjectives, and their volume or weight fractions.

However, composites can also have a (more interesting) second group of properties, which are not intrinsic to its constitutive phases: these are the so-called product tensor properties, first proposed by Van Suchtelen (1972, 1980). The appearance of new properties in a composite, which were not present in any individual phase that went into the composite, can be explained as follows. If one of the phases facilitates transformation of an applied independent variable A into an effect B, the inter-relationship of A and B can be characterized by the parameter $X = \partial B/\partial A$. Then, if the second phases converts the variable B to an effect C, the inter-relationship of B and C can be characterized by $Y = \partial C/\partial B$. Then, A must be able to be transformed into C, whose transformation can be characterized by a parameter that is the product of the component parameters: $\partial C/\partial A = (\partial C/\partial B)(\partial B/\partial A) = YX$. Thus, a composite of the two materials must posses a new property which can convert A into C: this property only results from the action of one phase on the other, and is absent in each phase when physically separated from the other.

It is fair to note that B.D.H. Tellegen had already in 1948 proposed a device based on ME composites, which was later named the Tellegen's gyrator (Tellegen 1948; Nan et al. 2008). Composites with strong ME effects were not known 60 years ago, and therefore Tellegen's conjecture ME device was not practically realized.

Finally, with regards to any composite materials, there are many possible ways to vary their physical properties, simply by changes and optimizations in the construction and dimensionality of the composite design.

The ME effect in composites is as result of elastically coupled piezoelectric and piezomagnetic effects. The mechanism of ME effects in hybrid composites is as follows: the piezomagnetic material is deformed under an applied magnetic field. This deformation results in a mechanical voltage that acts on the piezoelectric component, and hence induces an electric polarization change in the material via piezoelectricity. Obviously, the converse effect is also possible: an applied electric field causes the piezoelectric component to deform, resulting in mechanical voltage acting on the piezomagnetic material, generating a change in magnetization. Either directly or conversely, the net effect is a new tensor property of the composite—the ME effect: which consists of an electric polarization induced by an external magnetic field, and a magnetization induced by an external electric field.

Most of the known magnetically ordered materials have some measurable magnetostrictive effect: as it is a fourth rank polar tensor property, and thus must be exhibited by all crystal classes. However, the piezomagnetic effect is not required by symmetry to be present in all crystal classes; since it is a third rank axial tensor, its property matrix can become null by action of specific symmetry operations. As magnetostriction is the most prevalent form magnetoelastic coupling, deformations (ε_{ij}) induced by external magnetic fields (H) most frequently depend quadratically on field strength, rather than linearly: this is the definition of magnetostriction, i.e., $\varepsilon_{ij} = Q_{ijkl}H_kH_l$, where Q_{ijkl} is the magnetostriction coefficient. This fact makes use of composites difficult in device applications which require linearity over large field ranges. Linearization can only be achieved under applied dc bias magnetic field. In this case, the ME effect will be close to linear as long as the range of ac magnetic fields remains small in comparisons to that of the superimposed dc magnetic bias.

ME composites were prepared for the first time by van den Boomgard and coworkers. They used unidirectional solidification of a eutectic composition in the quinary system Fe–Co–Ti–Ba–O (Van den Boomgard et al. 1974; Van Run et al.

1974). This process promotes formation of alternating layers of magnetic spinel and piezoelectric perovskite phases. The unidirectional solidification process requires careful control of composition, in particular when one of the components (oxygen) is a gas. Investigations of said composites have shown that excess TiO_2 (1.5 weight %) allowed obtainment of a ME voltage coefficient of $\alpha_E = dE/dH = 50$ mV/(cmOe): which was considered quite high at that time. However, other compositions showed a much lower ME voltage coefficient in the range of 1–4 mV/(cmOe). In subsequent work, the authors reported a ME voltage coefficient of $\alpha_E = 30$ mV/(cmOe) for the eutectic composition $BaTiO_3$–$CoFe_2O_4$, again prepared by unidirectional solidification (Bichurin et al. 2007). This value was nearly an order of magnitude greater than that for the best single-phase response previously reported which was $\alpha_E = 20$ mV/(cmOe) for single crystal Cr_2O_3. Using the ME voltage coefficient, it is possible to obtain other relevant ME parameters, such as $\alpha = dP/dH = \alpha_E K \varepsilon_o$, where K is the average relative dielectric constant of the composite and ε_o the permittivity of free space. Using dielectric constants of $K = 500$ for the composite and $K = 11.9$ for Cr_2O_3, we can obtain an estimated value of $\alpha = 7.22 \times 10^{-10}$ s/m for the composite, which is approximately in 270 times higher than that of $\alpha = 2.67 \times 10^{-12}$ s/m for Cr_2O_3.

In ceramic composites, the value for the ME effect of $BaTiO_3$ and $NiFe_2O_4$ alloyed by cobalt and manganese was reduced relative to that prepared by unidirectional solidification. The maximum value of the ME voltage coefficient was 24 mV/(cmOe). The authors reported an unusual polarization behavior in which the field polarity was reversed at a Curie temperature (Van den Boomgard and Van Run 1976). Using the field created by volume charges in the composite allowed obtainment of make full poled samples. For ceramic composites of $BaTiO_3$ and $Ni(Co,Mn)Fe_2O_4$, it was found possible to obtain ME voltage coefficients of 80 mV/(cmOe).

There are other early studies of ceramic composites. Bunget and Raetchi observed a ME effect in Ni–Zn ferrite–PZT composites, and studied its dependence of applied magnetic field (Bunget and Raetchi 1981; Bunget and Raetchi 1982). The magnitude of the ME voltage coefficient was found to be 3.1 mV/(cmOe). Discussions of the potential for broad-band sensors based on $BaTiO_3$–$NiFe_2O_4$ composites with working frequencies up to 650 kHz should be noted. Composites of Mg–Mn ferrite–$BaTiO_3$ are simultaneously ferroelectric and ferrimagnetic: ferrite/$BaTiO_3$ composites with weight fractions of 30:70, 50:50, 70:30, and 90:10 have all revealed ferroelectric and ferromagnetic hysteresis. Subsequent measurements of the piezoelectric properties of ME composites have shown that the piezoelectric resonance frequency depends on applied magnetic field, where the maximum variation of resonance frequency was 0.2 % under $H = 875$ kA/m.

Magnetostrictive metals have, more recently, been investigated as alternatives to ferrites in ME laminated composites. Efforts have focused on use of permendur, Terfenol, and Metglas (Schmid et al. 1993; Bichurin 1997). These (Fe, Ni, Co) magnetostrictive alloys offer much larger magnetostrictions. The ME effect in layered composites based on these magnetostrictive alloys and piezoelectric PZT ceramics is significantly larger than those based on ferrites, which is important for engineering applications. The maximum ME voltage coefficient reported was about 4 V/(cmOe), which was obtained in multilayer structures of Terfenol and PZT (Zhai et al. 2008; Priya et al. 2007). This opens real possibilities for practical devices.

It is also relevant to note that these composites have been considered across a wide frequency bandwidth ranging from near dc to millimeter. Research on microwave ME effects in layered ferrite–piezoelectric composites has been performed (Bichurin et al. 1990). Application of external electric field to the piezoelectric phase is known to induce a shift in the FMR line frequency of the ferrite phase. An analogous effect was reported earlier in bulk ferrite–piezoelectric composites (Bichurin and Petrov 1988). Detailed analysis of the resonance ME effects in paramagnetic and magnetic ordered materials were performed (Bichurin et al. 1985). Microscopic theories of ME effects were presented for the magnetic resonance frequency range in magnetically ordered crystals with 3d-electrons. In addition, a theory for low-frequency ME effects and subsequent dispersion with increasing frequency have also been reported.

Finally, there are other important works and reviews that should be read. These include the following: (i) a theory of ME effects in homogeneous composites and heterogeneous structures (Getman 1994); (ii) applications of ME laminates in devices (Bichurin and Petrov 1998); and (iii) books and review articles (Fiebig 2005; Bichurin et al. 2007; Nan et al. 2008; Bichurin and Viehland 2012; Srinivasan 2010; Ederer and Nicola 2004; Eerenstein et al. 2006; Ramesh and Nicola 2007; Cheong and Mostovoy 2007) in which an analysis of the basic operational principles of ME composites is given.

1.2 Estimations of Composites' ME Parameters

Efforts Van den Boomgard et al. (1976) have been devoted to estimation of the ME voltage coefficient for composites, which were based on approximate models. Supposing that the permittivity of barium titanate exceeds considerably that of ferrite, and that the elastic stiffness of both phases are equal, the following relationship has been obtained

$$\begin{aligned}\alpha_E = (dE/dH)_{\text{composite}} &= (dx/dH)_{\text{composite}}(dE/dx)_{\text{composite}} \\ &= {}^{m}v(dx/dH)_{\text{ferrite}}(dE/dx)_{\text{BaTiO}_3};\end{aligned}$$

where dx/dH characterizes the material displacement under an applied H, dE/dx is its shape change under an applied E, and ${}^{m}v$ is the ferrite volume fraction.

Using optimistic values for the material parameters of $(dx/x)/dH \approx 6.28 \times 10^{-9}$ m/A, $dE/(dx/x) \approx 2 \times 10^{9}$ V/m, and ${}^{m}v = 0.5$, the maximum ME voltage coefficient was approximated to be $\alpha_E = 5$ V/(cmOe). A more precise estimate is as follows

$$\alpha_E = (dE/dH)_{\text{composite}} = {}^m v\,(dS/dH)_{\text{ferrite}}(l - {}^m v)\,(dE/dS)_{\text{piezoelectric}}.$$

By taking into account that $dE = dE_3 = g_{33}dT_3$ and $dS = (dT_3)/C_{33}$ (where g_{33} and C_{33} are the piezoelectric and stiffness coefficients of the piezoelectric phase, T the stress, and S the strain), the expression takes the following form:

$$\alpha_E = {}^m v\,(dS/dH)_{\text{ferrite}}(1 - {}^m v)\,(g_{33}C_{33})_{\text{piezoelectric}}.$$

Estimates for the ME voltage coefficient of $\alpha_E = 0.92$ V/(cmOe) have been obtained using this above formula.

To observe the ME effect in composites, it is necessary to apply a dc magnetic bias in direction of an applied ac magnetic field: this is because the primary magnetoelastic coupling mechanism is magnetostrictive, rather than piezomagnetism. A magnetic bias can be created by an attached permanent magnet, or by a nearby electromagnet. Bunget and Raetchi (1981, 1982) offered an alternative measurement method. In ME composites, since the electric polarization is a function of both changes in electric and magnetic fields, it is possible to apply them simultaneously and to then measure the polarization. The approach involves measuring the polarization under constant electric field, while varying the magnetic field. Then, the ratio of the change in polarization to the increment in magnetic field yields the ME sensitivity.

Newnham et al. (1978) offered a classification of composites based on dimensional connectivity types. For example, a composite with one of its phase connected in all three directions (denoted by an index 3) and with a second phase that is isolated having connectivity in no direction (denoted by index 0) was designated as a composite with connectivity type of 3–0. In ME composites containing magnetostrictive ferrites, the ferrite phase has considerably smaller resistance than the piezoelectric one. This leads to a strong dependence of the composite resistance on phase connectivity: the highest resistance occurs for series connection of composite components, whereas the lowest one occurs for parallel. Some ferrites are semiconductors, whose resistance strongly decreases with increasing temperature. To observe ME effects in a composite, it is necessary to permanently pole them under electric field, to get them to demonstrate a piezoelectric effect. However, in ferrites with notable conductivity, it is difficult to get the bulk resistance of the composite sufficiently high to be able to achieve good poling in the piezoelectric phases. However, using a composite with a 0–3 connectivity between ferrite and piezoelectric phases allows for notable increases in the bulk resistivity. Because of this fact, 0–3 composites of ferrite and piezoelectric phases are easier to pole. Consequently, they offer a better practical connectivity form for ME composites made of these materials systems.

Laminate composites, which to date are of most interest for giant magnetoelectricity, have a 2–2 connectivity type: in which case, layers of one phase are stacked on another, in an alternative manner. In this case, each phase has connectivity in two directions in the layer plane, but is not connected with other layer of the same phase which are separated by those of the second phase. This composite type consists of layers that are mechanically connected in series, but which can be electrically connected in either series or parallel as the electrodes can be configured in different manners. Harshe et al. (1993a, b) reported calculations of the ME voltage coefficient for composites with a 2–2 connectivity type. They determined that the ME voltage coefficient was the ratio of the electric field induced across the piezoelectric phase to that of the magnetic field applied to magnetostrictive one: i.e., $\alpha_E = {}^pE/{}^mH$.

Figure 1.1 shows various alternative models of ME bilayers with different boundary conditions. The first variant corresponds to a layered composite with ideal mechanical connection between layers. The second bilayer composite variant has boundary conditions representing a clamping of magnetostrictive and piezoelectric layers at both ends, with thin layers of lubricant between all surfaces to minimize friction. The third considers magnetostrictive and piezoelectric layers that are rigidly clamped at both ends. And, the fourth illustrates clamping at both ends, with thin lubricant layers between clamp and sample surfaces to minimize friction. We assume in all four cases, that both phases are perfectly bonded together.

There are some special considerations for the modeling of ME composites that need to be mentioned. First, piezoelectric layers of a composite can be electrically connected in series or parallel. Depending upon which piezoelectric d_{33} or d_{31} and magnetostriction $\lambda_{//}$ or $\lambda_{\perp}$ coefficients that one wishes to take advantage of, different multilayer ME composite configurations can be created (Harshe et al. 1993). Investigations have been devoted to longitudinal ME effects in bilayer composite based on Terfenol-D (Mori and Wuttig 2002), in which case a ME voltage coefficient of 1.43 V/cmOe was reported. Second, optimally poled piezoelectric layer effects can be achieved in multilayer structures where ferrite layers are paralleled by intermediate electrodes. In this case, we get a composite with a mechanical connectivity type of 2–2, but with an electrical connectivity type of 3–0. This can be considered as a composite with a mixed connectivity type.

It is often required to predict the effective composite parameters from their constitutive components. Such effective parameters can be determined by the Maxwell-Garnett equation 47. For example, in the case of a 0–3 composite, the effective permittivity of a continuous dielectric matrix having a permittivity ε_1 containing isolated second phase particles with ε_2 can be given as follows

$$\varepsilon_{\ni\phi\phi} = \varepsilon_1[2\varepsilon_1 + \varepsilon_2 + 2y(\varepsilon_1 - \varepsilon_2)]/[2\varepsilon_1 + \varepsilon_2 - y(\varepsilon_1 - \varepsilon_2)]; \quad (1.14)$$

where y is the volume fraction of any additional component.

Calculations of the ME properties of composites have some discrepancy with measured values. Measurement of the ME effect in sintered composites of $NiFe_2O_4$ or $CoFe_2O_4$ and $BaTiO_3$ have been reported (Van den Boomgard et al. 1974). Samples in the form of thin disks were polarized by an electric field applied perpendicular to the sample plane. The ME coefficient was then measured for two cases: (i) transverse fields, where dc and ac magnetic fields were parallel to each other and to the plane of the disk (along directions 1 and 2), which were

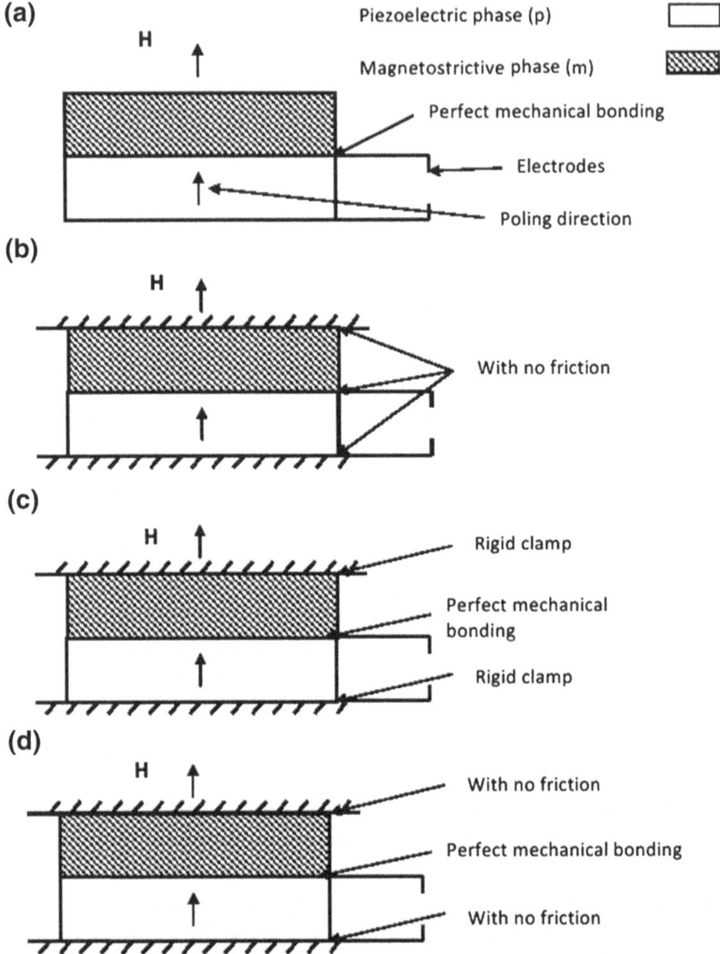

Fig. 1.1 Different models of ME bilayer. **a** The first variant corresponds to a layered composite with ideal mechanical connection between layers. **b** The second bilayer composite variant has boundary conditions representing a clamping of magnetostrictive and piezoelectric layers at both ends, with thin layers of lubricant between all surfaces to minimize friction. **c** The third considers magnetostrictive and piezoelectric layers that are rigidly clamped at both ends. **d** The fourth illustrates clamping at both ends, with thin lubricant layers between clamp and sample surfaces to minimize friction.

perpendicular to an ac electric field (along direction 3); and (ii) longitudinal fields, where all three fields (dc and ac magnetic, and ac electric) were parallel to each other and perpendicular to the sample's plane. In general, bulk ceramic composites exhibit values of the ME voltage coefficient that are notably lower than those predicted theoretically by continuum mechanics (Bichurin et al. 2002). One reason for this is the low resistivity of the ferrite phase that was aforementioned, that (i) decreases the electric field which can be applied during poling of the piezoelectric phase, resulting in insufficient insufficient polarization; and (ii) leakage current across the electrodes of the composites, which results in an inability of the composite to maintain charge after it has been induced by applied magnetic fields via the piezoelectric effect. The essential advantage of bulk composites is the possibility of reaching the required values of the given parameters using combinations of the constituent phases that have the necessary values of electric and magnetic materials parameters, and alternatively by tuning the effective composite parameters by adjusting the relative phase volume fractions.

Cubic models of ferrite–piezoelectric ME composites with a 3–0 and 0–3 connectivity types have been considered in Harshe et al. (1993), which allow for numerical computation of ME coefficients. However, this early theoretical model was inadequate: as evidenced by experimental observations that have shown ME voltage coefficients (8×10^{-3} V/(cmOe)) more than two orders of magnitude lower than theoretical predicted one (3.9 V/(cmOe)). Ma et al. (2011) developed a computational approach to ME effects in bulk composites, based on a Green's functional approach and perturbation theory. The work offered the perspective of a three-phase composite with high mechanical and ME properties.

The ME effect could find wide applications in various types of electronic devices. Possible applications were considered earlier in Smolenskii and Chupis (1982), Fiebig et al. (2004), Bichurin et al. (1990, 2002a, b), Bichurin and Petrov (1998), microwave phase shifters were considered in (Bichurin et al. 2002c), ME magnetic field and microwave power sensors in (Tatarenko et al. 2010). For such applications, ME composites are necessary (rather than single-phase crystals) due to their high values of ME coefficients and higher working temperature range. However, application of ME composites is hindered by bad reproducibility of effective composite parameters. For example, good mechanical connection across layer interface between phases is necessary to achieve good ME coupling. Furthermore, the constituent phases of a composite should not react with each other. This is a concern in particular for sintered composites based on ferrite–piezoelectric ceramics, as very high temperatures are used in the densification process: avoidance of chemical reactions at interfacial areas can then complicate processing of sintered composites.

1.3 Direct and Converse Effects

Magnetoelectric susceptibility is known to be the fundamental parameter describing the ME coupling (Zhai et al. 2007). Recently, it was shown that direct and converse ME susceptibilities are equivalent for strain-coupled two-phase systems (Lou et al. 2012), i.e.

$$\alpha = \frac{\partial P}{\partial H} = \mu_0 \frac{\partial M}{\partial E} \tag{1.15}$$

ME coefficients for direct and converse ME effects can be expressed in terms of ME susceptibility. Thus, ME voltage coefficient can be defined as $\alpha_E = \frac{\partial E}{\partial H}$ under open electric circuit condition and can be expressed as $\alpha_E = \frac{\alpha}{\varepsilon}$ where ε is effective permittivity. Similarly, electrically induced magnetic field can be determined as follows: $\alpha_H = \frac{\partial H}{\partial E} = \frac{\alpha}{\mu}$ under open magnetic circuit condition where μ is effective permeability.

Direct and converse coupling strengths may be simply related for two-phase systems via the Maxwell relation, provided it is rewritten as $dp/dH = \mu_O\, dm/dE$, in order to describe the entire system via effective magnetic dipole moment m and effective electrical dipole moment p.

For describing the ME response due to mechanical-electric-magnetic coupling, the constitutive equation for the piezoelectric effect can be written in the following form:

$$ {}^{p}S_i = {}^{p}s_{ij}pT_j + {}^{p}d_{ki}E_k, \tag{1.16}$$

$$ D_k = {}^{p}d_{ki}{}^{p}T_i + {}^{p}\varepsilon_{kn}E_n, \tag{1.17}$$

where ${}^{p}S_i$ is astrain tensor component of the piezoelectric phase; E_k is a vector component of the electric field; D_k is a vector component of the electric displacement; ${}^{p}T_i$ is astress tensor component of the piezoelectric phase; ${}^{p}s_{ij}$ is a compliance coefficient of the piezoelectric phase; ${}^{p}d_{ki}$ is a piezoelectric coefficient of the piezoelectric phase; and ${}^{p}\varepsilon_{kn}$ is a permittivity matrix of the piezoelectric phase.

Analogously, the strain and magnetic induction tensors of the magnetostrictive phase are given as:

$$ {}^{m}S_i = {}^{m}s_{ij}{}^{m}T_j + {}^{m}q_{ki}H_k, \tag{1.18}$$

$$ B_k = {}^{m}q_{ki}{}^{m}T_i + {}^{m}\mu_{kn}H_n, \tag{1.19}$$

where ${}^{m}S_i$ is a strain tensor component of the magnetostrictive phase; ${}^{m}T_j$ is a stress tensor component of the magnetostrictive phase; ${}^{m}s_{ij}$ is a compliance coefficient of the magnetostrictive phase; H_k is a vector component of magnetic field; B_k is a vector component of magnetic induction; ${}^{m}q_{ki}$ is a piezomagnetic coefficient; and ${}^{m}\mu_{kn}$ is a permeability matrix.

In what follows, we consider the trilayer magnetoelectricstructure (Metglas/PZT/Metglas) in the form of thin and narrow strip so that 1D theoretical model can be used. In-plane magnetic and out-of plane electric fields are supposed to be directed along 1 and 3 axis, correspondingly. The following boundary conditions can be used to derive expressions for ME coefficients.

$$ \begin{aligned} {}^{p}S_1 &= {}^{m}S_1; \\ {}^{p}T_1{}^{p}t &= -{}^{m}T_1^{m}t; \end{aligned} \tag{1.20}$$

where ${}^{p}t$ and ${}^{m}t$ denote thicknesses of the piezoelectric and magnetostrictive phases, respectively.Stress components can be found from (1.20) by taking into account (1.16) and (1.18). Substituting the found expressions for stress components into (1.17) and (1.19) enables one to express the electric and magnetic inductions in terms of electric and magnetic fields. Then ME susceptibility can be found as $\partial D_3/\partial H_1$ or $\partial B_1/\partial E_3$.

$$\alpha_{13} = \frac{{}^{m}q_{11}{}^{p}d_{31}}{{}^{p}s_{11} + r{}^{m}s_{11}}, \qquad (1.21)$$

where $r = {}^{p}t/{}^{m}t$ with ${}^{p}t$ and ${}^{m}t$ denoting the thicknesses of piezoelectric and piezomagnetic layers, correspondingly.

For DME, the transverse ME voltage coefficient can be found as the ratio of induced electric field to applied magnetic field under open electric circuit condition, $D_3 = 0$.

$$\alpha_{E31} = -\frac{{}^{m}q_{11}{}^{p}d_{31}}{[(1 - {}^{p}K_{31}^{2}){}^{p}s_{11} + r{}^{m}s_{11}]{}^{p}\varepsilon_{33}}. \qquad (1.22)$$

For CME at transverse field orientation, the relevant ME coefficient is defined as the ratio of induced magnetic field to applied electric field at open magnetic circuit condition, $B_1 = 0$.

$$\alpha_{H13} = -\frac{{}^{m}q_{11}{}^{p}d_{31}}{[(1 - {}^{m}K_{11}^{2}){}^{m}s_{11} + p{}^{p}s_{11}]{}^{m}\mu_{11}}, \qquad (1.23)$$

where $p = {}^{m}t/{}^{p}t$.

Experimentally, ME susceptibility is obtained under closed electric circuit condition for DME (output voltage equals zero and output current should be measured) and under closed magnetic circuit condition for CME. As opposed to ME susceptibility, ME voltage coefficient can be measured at open electric circuit condition. Similarly, α_H should be measured at open magnetic circuit condition.

Since the ME effect in layered structures is due to mechanically coupled piezoelectric and magnetostrictive subsystems, it sharply increases in the vicinity of the EMR. Mechanical oscillations of a ME composite can be induced either by alternating magnetic or electric fields. The equation of medium's axial motion can be written as:

$$\bar{\rho}\frac{\partial^2 u_i}{\partial t^2} = V\frac{\partial^p T_{ij}}{\partial x_j} + (1 - V)\frac{\partial^m T_{ij}}{\partial x_j}, \qquad (1.24)$$

where u_i is the displacement vector component, $\bar{\rho} = ({}^{p}t^{p}\rho + {}^{m}t^{m}\rho)/({}^{p}t + {}^{m}t)$ is the average mass density, ${}^{p}\rho$ and ${}^{m}\rho$ are the densities of ferroelectric and ferromagnet. If the length of the electromagnetic wave exceeds the spatial size of the composite by some orders of magnitude, then it is possible to neglect gradients of the electric and magnetic fields within the sample volume. Therefore, based on elastodynamics, magnetostatics and electrostatics, the equation of medium's axial motion for harmonic vibrationsare governed by:

$$\frac{\partial^2 u_1}{\partial x^2} = -k^2 u_1, \tag{1.25}$$

where $k = \omega\sqrt{({}^p\rho^p t + {}^m\rho^m t)\left(\frac{{}^p t}{{}^p s} + \frac{{}^m t}{{}^m s^B}\right)^{-1}}$ with ${}^p s$ and ${}^m s^B$ denoting the thicknesses of piezoelectric layer at constant electric field and piezomagnetic layer at constant magnetic induction, correspondingly. Equation (1.25) should be solved for the boundary conditions that consist of vanishing total force at the ends of trilayer.

Solving (1.25) for u_l and substituting the solution into (1.16) enables us to obtain the stress component. Once the stress components are determined, the expression for the ME susceptibility can be obtained as $\frac{\partial \bar{D}_3}{\partial H_1}$ with $\bar{D}_3$ denoting the average electric induction:

$$\alpha_{13} = \frac{2{}^m q_{11}{}^p d_{31}\ \tan({}^{kL}\!/\!_2)}{[(1 + {}^m K_{11}^2)(p^p s_{11} + {}^m s_{11}^B)]kL - 2{}^m K_{11}^2 p^p s_{11}\ \tan({}^{kL}\!/\!_2)}$$

where L is the sample length,

It should be noted that an increase in ME susceptibility occurs at resonance frequency for both DME and CME. To obtain the expression for the ME voltage coefficient for DME, we use the open circuit condition, $\int_{-L/2}^{L/2} D_3 dx = 0.$

$$\alpha_{E31} = \frac{2{}^m q_{11}{}^{p}d_{31}/{}^p\varepsilon_{33}\ \tan({}^{kL}\!/\!_2)}{[2{}^m K_{11}^2 {}^p s_{11} - 2{}^p K^{2m} s_{11}^B r]\tan({}^{kL}\!/\!_2) - ({}^p s_{11} + r^m s_{11}^B)kL}$$

Increase in ME voltage coefficient is obtained at antiresonance frequency.

1.4 Conclusions

In this chapter, we discussed the ME properties of ferrite–piezoelectric composites, to create new ME composites with enhanced ME couplings that would enable them for application in functional electronics devices. To address this important scientific and technical goal, a generalization of various theoretical and experimental studies of ME composites has been given. One of the main tasks according to the formulated approach is a comparative analysis of ME composites that have different connectivity types. The relative simplicity of manufacturing multilayer composites with a 2–2 type connectivity having giant ME responses is an important benefit. In addition, composites with 3–0 and 0–3 connectivity types can also be made in considerable quantity by a minimum monitoring of the synthesis process.

Any material with a high piezoelectric constant is a good choice for the piezoelectric phase in ME composites. The most suitable ones are $Pb(Zr,Ti)O_3$, $BaTiO_3$, and $Pb(Mg_{1/3}Nb_{2/3})O_3$–$PbTiO_3$, which is due to their large piezoelectric coefficients. Analogously, any material with a high piezomagnetic coefficient at low magnetic biases is a good candidate for the magnetostrictive phase. The most suitable choices are ferrites and ferromagnetic metals (such as Terfenol-D and Metglas).

Availability of theoretical models for composite properties is necessary to interpret experimental data and to restrict oneself among the multitude of composite configurations. Theoretical estimations of the ME voltage coefficient for series and parallel composite models, and also a cubic model for composites with a 3–0 connectivity type (Tatarenko et al. 2010; Bichurin et al. 2002d), are known. However, as already mentioned, ME voltage coefficient was computed as the ratio of an electric field induced in the piezoelectric phase to the magnetic field applied to the magnetic one: i.e., $\alpha_E = {}^pE/{}^mH$. But, in reality, the internal fields in composite components can be significantly different from the external fields. In particular, formulas (Harshe et al. 1993a, b) predicted a maximum ME voltage coefficient in a pure piezomagnetic phase (i.e., ${}^pv = 0$): which distinctly mismatches reality. In addition, generalized models for composite based on an effective medium method were presented. These offer determination of the effective composite parameters with phase connectivity types of 2–2, 3–0, and 0–3 that are based on exact solutions. The ultimate purpose of any theoretical work must be to predict the ME parameters—both susceptibility and voltage coefficients—as these are the most basic parameters of magnetoelectricity.

It is important to realize that ME effects occur over a broad frequency bandwidth, extending from the quasi-static to millimeter ranges. This offers important opportunities in potential device applications. It makes possible new concepts in sensing, gyrators, microwave communications, phase shifters, just to name a few. It also complicates the understanding of magnetoelectricity, as there are significant changes in its spectra with frequency. There are strong enhancements in the ME coefficients near both the electromechanical and magneto mechanical resonances. In addition, there is the important problem of studying dispersion in the ME parameters over a broad frequency range of $10^{-3} < f < 10^{10}$ Hz. Relaxation parameters depend on connectivity type, composite geometry and structure, and volume fraction of constituent phases.

This book surveys the ME effect in ME composites over a wide frequency range, offering suggestions for making new ME materials with sufficient exchange to enable practical applications. Generalized theoretical and experimental studies will be presented that try to gain advantage by comparing existing solutions with existing data.

References

Asher E (1969) The interaction between magnetization and polarization: phenomenological symmetry consideration. J Phys Soc Jpn 28:7

Astrov DN (1961) Magnetoelectric effect in chromium oxide. Sov Phys JETP 13:729

Aubert G (1982) A novel approach of the magnetoelectric effect in antiferromagnets. J Appl Phys 53:8125

Bichurin MI (ed) (1997) In: Proceedings of the 3rd international conference on magnetoelectric interaction phenomena in crystals (MEIPIC-3, Novgorod). Ferroelectrics, vol 204, p 356

Bichurin MI (ed) (2002) In: Proceedings of the fourth conference on magnetoelectric interaction phenomena in crystals (MEIPIC-4, Veliky Novgorod). Ferroelectrics, vol 279–280, p 386

Bichurin MI, Petrov VM (1987) Influence of electric field on antiferromagnetic resonance spectrum in iron borate. Phys Solid State 29:2509

Bichurin MI, Petrov VM (1988) Magnetic resonance in layered ferrite-ferrielectric structures. Sov Phys JETP 58:2277

Bichurin MI, Petrov VM (1998) Magnetoelectric materials in the microwave range. Yaroslav-the-Wise Novgorod State University, Novgorod, p 154 (in Russian)

Bichurin MI, Viehland D (eds) (2012) Magnetoelectricity in composites. Pan Stanford Publshing, Singapore, 273 p

Bichurin MI, Didkovskaya OS, Petrov VM, Sofronev SE (1985) Resonant magnetoelectric effect in composite materials. Izv Vuzov Ser Physic 1:121 (in Russian)

Bichurin MI, Venevtsev YN, Didkovskaya OS, Petrov VM, Fomich NN (1990) Magnetoelectric materials: technology features and application perspectives. In: Magnetoelectric substances, Nauka, Moscow pp 118–132 (in Russian)

Bichurin MI, Petrov VM, Petrov RV, Bukashev FI, Smirnov AY (2002a) Electrodynamic analysis of strip line on magnetoelectric substrate. Ferroelectrics 280:203–209

Bichurin MI, Petrov VM, Petrov RV, Kiliba YV, Bukashev FI, Smirnov AY (2002b) Magnetoelectric sensor of magnetic field. Ferroelectrics 280:199–202

Bichurin MI, Petrov VM, Petrov RV, Kapralov GN, Kiliba YV, Bukashev FI (2002c) Magnetoelectric microwave devices. Ferroelectrics 280:211–218

Bichurin MI, Petrov VM, Srinivasan G (2002d) Modelling of magnetoelectric effect in ferromagnetic/piezoelectric multilayer composites. Ferroelectrics 280:165

Bichurin MI, Viehland D, Srinivasan G (2007) Magnetoelectric interactions in ferromagnetic—piezoelectric layered structures: phenomena and devices. J Electroceram 19:243

Brown Jr WF, et al (1968) Upper bound on the magnetoelectric susceptibility. Phys Rev 168:574

Bunget I, Raetchi V (1981) Magnetoelectric effect in the heterogeneous system NiZn ferrite—PZT ceramic. Phys Stat Sol 63:55

Bunget I, Raetchi V (1982) Dynamic magnetoelectric effect in the composite system of NiZn ferrite and PZT ceramics. Rev Roum Phys 27:401

Cheong S-W, Mostovoy M (2007) Multiferroics: a magnetic twist for ferroelectricity. Nat Mater 6:13–20

Dzyaloshinskii IE (1960) On the magneto-electrical effect in antiferromagnets. Sov Phys JETP 10:628

Ederer C, Spaldin NA (2004) Magnetoelectrics: a new route to magnetic ferroelectrics. Nat Mater 3:849–851

Eerenstein W, Mathur ND, Scott JF (2006) Multiferroic and magnetoelectric materials. Nature 442:759–765

Fiebig M (2005) Revival of the magnetoelectric effect. J Phys D Appl Phys 38:R1

Fiebig M, Eremenko VV, Chupis IE (eds) (2004) In: Proceedings of the fifth conference on magnetoelectric interaction phenomena in crystals (MEIPIC-5, Sudak) Kiuwer Academic Publishers, NATO Sciences Series, 334 p

Folen VJ, Rado GT, Stalder EW (1961) Anysotropy of the magnetoelectric effect in Cr_2O_3. Phys Rev Lett 6:607

Freeman AJ, Schmid H (1975) Magnetoelectric interaction phenomena in crystals. Gordon and Breach, London, 228 p

Getman I (1994) Magnetoelectric composite materials: theoretical approach to determine their properties. Ferroelectrics 162:45−50

Harshe G, Dougherty JO, Newnham RE (1993a) Theoretical modelling of multilayer magnetoelectric composites. Int J Appl Electromagn Mater 4:145

Harshe G, Dougherty JP, Newnham RE (1993b) Theoretical modelling of 3-0, 0-3 magnetoelectric composites. Int J Appl Electromagn Mater 4:161

Landau LD, Lifshitz EM (1980) Statistical physics, 3rd edn. Pergamon Press, Oxford, 562 p

Lou J, Pellegrini GN, Liu M, Mathur ND, Sun NX (2012) Inequivalence of direct and converse magnetoelectric coupling at electromechanical resonance. Appl Phys Lett 100:102907

Ma J, Hu J, Li Z, Nan C-W (2011) Recent progress in multiferroic magnetoelectric composites: from bulk to thin films. Adv Mater 23:1062–1087

Mori K, Wuttig M (2002) Magnetoelectric coupling in terfenol-d/polyvinylidenedifluoride composites. Appl Phys Lett 81:100

Nan C-W, Bichurin MI, Dong S, Viehland D, Srinivasan G (2008) Multiferroic magnetoelectric composites: historical perspectives, status, and future directions. J Appl Phys 103:031101

Newnham RE, Skinner DP, Cross LE (1978) Connectivity and piezoelectric-pyroelectric composites. Mater Res Bull 13:525

O'Dell TH (1970) The electrodynamics of magnetoelectric media. North-Holland Publication Company, Amsterdam, 304 p

Opechovski W (1975) Magnetoelectric symmetry. In: Freeman A, Schmid H (eds) Proceedings of symposium on magnetoelectric interaction in crystals, USA, 1973. Gordon and Breach Science Publication, New York, p 47

Priya S, Islam R, Dong S, Viehland D (2007) Recent advancements in magnetoelectric particulate and laminate composites. J Electroceram 19:149–166

Ramesh R, Nicola A (2007) Spaldin, Multiferroics: progress and prospects in thin films. Nat Mater 6:21–29

Rivera J-P (2009) A short review of the magnetoelectric effect and related experimental techniques on single phase (multi-) ferroics. Eur Phys J B 71:299–313

Santoro RP, Newnham RE (1966) Survey of magnetoelectric materials. Technical Report AFML TR-66-327, Air Force Materials Lab, Ohio

Schmid H, Janner A, Grimmer H, Rivera J-P, Ye Z-G (eds) (1993) In: Proceedings of the 2nd international conference on magnetoelectnc interaction phenomena in crystals (MEIPIC-2, Ascona) Ferroelectrics, vol 161–162, 748 p

Smolenskii GA, Chupis IE (1982) Ferroelectromagnets. Sov Phys Usp 25:475–493

Srinivasan G (2010) Magnetoelectric composites. Annu Rev Mater Res 40:153

Tatarenko AS, Bichurin MI, V.Gheevarughese, et al (2010) Microwave magnetoelectric effects in ferrite-piezoelectric composites and dual electric and magnetic field tunable filters. J Electroceram 24:5

Tellegen BDH (1948) The gyrator, a new electric network element. Philips Res Rep 3:81

Van den Boomgard J, Van Run AMJG (1976) Poling of a ferroelectric medium by means of a built-in space charge field with special reference to sintered magnetoelectric composites. Solid State Commun 19:405

Van den Boomgard J et al (1974) An in situ grown eutectic magnetoelectric composite materials: part I. J Mater Sci 9:1705

Van den Boomgard J, Van Run AMJG, Van Suchtelen J (1976) Magnetoelectricity in piezoelectric-magnetostrictive composites. Ferroelectrics 10:295

Van Run AMJG et al (1974) An in situ grown eutectic magnetoelectric composite materials: part II. J Mater Sci 9:1710

Van Suchtelen J (1972) Product properties: a new application of composite materials. Philips Res Rep 27:28

Van Suchtelen J (1980) Non structural application of composite materials. Ann Chim Fr 5:139

Wang KF, Liu J-M, Renc ZF (2009) Multiferroicity: the coupling between magnetic and polarization orders. Adv Phys 58:321

Zhai J, Li J, Viehland D, Bichurin MI (2007) Large Magnetoelectric susceptibility: The fundamental property of piezoelectric and magnetostrictive laminated composites. J Appl Phys 101:014102

Zhai J, Xing Z, Dong S, Li J, Viehland D (2008) Magnetoelectric laminate composites: an overview. J Am Ceram Soc 91:351–358

Chapter 2
Low-Frequency Magnetoelectric Effects in Magnetostrictive-Piezoelectric Composites

Abstract In this chapter, we discuss the theoretical modeling of low-frequency ME effect in layered and bulk composites based on magnetostrictive and piezoelectric materials. Our analysis rests on the effective-medium approach and exact calculation based on elastostatic, electrostatic and magnetostatic equations. The expressions for effective parameters including ME susceptibilities and ME voltage coefficients as functions of material parameters and volume fractions of components are obtained. Longitudinal, transverse and in-plane field orientations are considered. The use of the offered model has allowed to estimate the ME effect in ferrite cobalt–barium titanate, ferrite cobalt–PZT, ferrite nickel–PZT, lanthanum-strontium manganite–PZT composites adequately.

2.1 Symmetric Layered Structures

Using layered structures enables one to overcome a series of difficulties that are characteristic for bulk composites. The reasons for the giant ME effects in layered composites are: (a) high piezoelectric and piezomagnetic coefficients in individual layers, (b) effective stress transfer between layers, (c) ease of poling and subsequent achievement of a full piezoelectric effect, and (d) ability to hold charge due to suppression of leakage currents across composites with a 2–2 connectivity.

Prior theoretical models based on mechanics and constitutive relationships by Harshe et al. (1993) were restricted to account for longitudinal ME voltage coefficient in laminates having ideal mechanical connection at the interfaces between layers. Principal disadvantages of this earlier approach were as follows: (i) For the case of longitudinally oriented fields, the effect of the magnetic permeability of the ferrite phase was ignored. Diminution of interior (local) magnetic fields results in a weakening of ME interactions via demagnetization fields. (ii) The case of fields applied in cross orientations to the ME layer connectivity was not considered, which later experimental investigations revealed large ME responses.

M. Bichurin and V. Petrov, *Modeling of Magnetoelectric Effects in Composites*, Springer Series in Materials Science 201, DOI: 10.1007/978-94-017-9156-4_2,

In this work, we present a summary of a more recent theory of ME laminate composites, which are free from the disadvantages mentioned just above. The approach is based on continuum mechanics, and considers the composite as a homogeneous medium having piezoelectric and magnetostrictive subsystems. To derive the effective material parameters of composites, an averaging method consisting of two steps (Bichurin et al. 2002a, b, 2003, 2009; Gheevarughese et al. 2007) should be used. In the first step, the composite is considered as a structure whose magnetostrictive and piezoelectric phases are distinct and separable. ME composites are characterized by the presence of magnetic and electric subsystems interacting with each other.

The constitutive equation for the piezoelectric effect can be given in the following form:

$$^{p}S_i = {}^{p}s_{ij}{}^{p}T_j + {}^{p}d_{ki}{}^{p}E_k, \tag{2.1}$$

$$^{p}D_k = {}^{p}d_{ki}{}^{p}T_i + {}^{p}\varepsilon_{kn}{}^{p}E_n; \tag{2.2}$$

where $^{p}S_i$ is a strain tensor component of the piezoelectric phase; $^{p}E_k$ is a vector component of the electric field; $^{p}D_k$ is a vector component of the electric displacement; $^{p}T_i$ is a stress tensor component of the piezoelectric phase; $^{p}s_{ij}$ is a compliance coefficient of the piezoelectric phase; $^{p}d_{ki}$ is a piezoelectric coefficient of the piezoelectric phase; and $^{p}\varepsilon_{kn}$ is a permittivity matrix of the piezoelectric phase.

Analogously, the strain and magnetic induction tensors of the magnetostrictive phase are respectively

$$^{m}S_i = {}^{m}s_{ij}^{m}T_j + {}^{m}q_{ki}{}^{m}H_k, \tag{2.3}$$

$$^{m}B_k = {}^{m}q_{ki}{}^{m}T_i + {}^{m}\mu_{kn}{}^{m}H_n, \tag{2.4}$$

where $^{m}S_i$ is a strain tensor component of the magnetostrictive phase; $^{m}T_j$ is a stress tensor component of the magnetostrictive phase; $^{m}s_{ij}$ is a compliance coefficient of the magnetostrictive phase; $^{m}H_k$ is a vector component of magnetic field; $^{m}B_k$ is a vector component of magnetic induction; $^{m}q_{ki}$ is a piezomagnetic coefficient; and $^{m}\mu_{kn}$ is a permeability matrix.

Prior models assumed that the connection at interfaces between layers was ideal. However, in this chapter, we assume that there is a coupling parameter $k = ({}^{p}S_i - {}^{p}S_{i0})/({}^{m}S_i - {}^{p}S_{i0})$ $(i = 1, 2)$, where $^{p}S_{i0}$ is a strain tensor component assuming no friction between layers (Bichurin et al. 2003). This interphase–interface elastic–elastic coupling parameter depends on interface quality, and is a measure of a differential deformation between piezoelectric and magnetostrictive layers. The coupling parameter is $k = 1$ for the case of an ideal interface, and *is* $k = 0$ for the case of no friction.

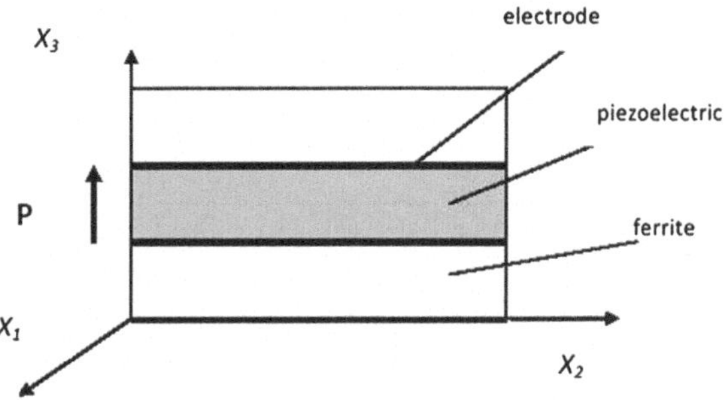

Fig. 2.1 Schematic of the layered composite structure

In the second step of the averaging method to derive the effective materials parameters, the bilayer composite is considered as a homogeneous solid, whose behavior can be described by the following coupled sets of linear algebraic equations:

$$\begin{aligned} S_i &= s_{ij}T_j + d_{ki}E_k + q_{ki}H_k, \\ D_k &= d_{ki}T_i + \varepsilon_{kn}E_n + \alpha_{kn}H_n, \\ B_k &= q_{ki}T_i + \alpha_{kn}E_n + \mu_{kn}H_n; \end{aligned} \tag{2.5}$$

where S_i is a strain tensor component; T_j is a stress tensor component; E_k is a vector component of the electric field; D_k is a vector component of the electric displacement; H_k is a vector component of the magnetic field; B_k is a vector component of the magnetic induction; s_{ij} is an effective compliance coefficient; d_{ki} is a piezoelectric coefficient; q_{ki} is a piezomagnetic coefficient; ε_{kn} is an effective permittivity; μ_{kn} is a permeability coefficient; and α_{kn} is a ME coefficient.

The simultaneous solution of the coupled sets of linear algebraic equations allows one to find the effective parameters of a composite.

Let us consider that the layers of a composite are oriented along the planes (X_1, X_2), and that the direction X_3 is perpendicular to the same plane. In this case, the direction of polarization in a sample coincides with the X_3 axis. If we by choice apply a constant magnetic bias and variable magnetic field along the same direction coincidental with that of the polarization, then any resultant electric field will also be parallel to the X_3 axis, as shown in Fig. 2.1. This summarization supposes that the symmetry of the piezoelectric phase is ∞m, and that of the magnetic phase is cubic. The following boundary conditions can then be used to derive expressions for ME coefficients.

$$\begin{aligned} {}^pS_i &= k^mS_i + (1-k)^pS_{i0}; \quad (i = 1, 2) \\ {}^pT_i &= -{}^mT_i(1-v)/v; \quad (i = 1, 2) \end{aligned} \tag{2.6}$$

where $v = pv/(pv + {}^mv)$, ${}^pv$ and ${}^mv$ denote the Poisson's ratio of the piezoelectric and magnetostrictive phases, respectively; and pS_{10} and pS_{20} are the strain tensor components for $k = 0$.

Using continuity conditions for magnetic and electric fields, and using open circuit condition, one can then obtain the following expressions for the ME susceptibility, and longitudinal ME voltage coefficient.

$$\alpha_{33} = 2\frac{k\mu_0(v-1)^p d_{31}{}^m q_{31}}{[\mu_0(v-1) - {}^m\mu_{33}v][kv({}^m s_{12} + {}^m s_{11}) - ({}^p s_{11} + {}^p s_{12})(v-1)] + 2^m q_{31}^2 kv^2};$$

$$\alpha_{E,33} = \frac{E_3}{H_3} = 2\frac{\mu_0 kv(1-v)^p d_{31}{}^m q_{31}}{\{2^p d_{31}^2(1-v) + {}^p\varepsilon_{33}[({}^p s_{11} + {}^p s_{12})(v-1) - v({}^m s_{11} + {}^m s_{12})]\}}$$
$$\times \frac{[({}^p s_{11} + {}^p s_{12})(v-1) - kv({}^m s_{11} + {}^m s_{12})]}{\{[\mu_0(v-1) - {}^m\mu_{33}v][kv({}^m s_{12} + {}^m s_{11}) - ({}^p s_{11} + {}^p s_{12})(v-1)] + 2^m q_{31}^2 kv^2\}} \tag{2.7}$$

The earlier expression obtained by Harshe et al. (1993) matched to our theory for the special case of $k = 1$, provided that the magnetic field is applied only to the ferrite phase.

The model presented above allows for the determination of the longitudinal ME coefficients as functions of volume fractions, physical parameters of phases, and elastic–elastic interfacial coupling parameter k.

Next we consider the transverse fields' orientation that corresponds to E and δE being applied along the X_3 direction, and H and δH along the X_1 direction (in the sample plane). The expressions for the ME susceptibility, and transverse ME voltage coefficient are then respectively

$$\alpha_{31} = \frac{(v-1)v({}^m q_{11} + {}^m q_{21})^p d_{31} k}{(v-1)({}^p s_{11} + {}^p s_{12}) - kv({}^m s_{11} + {}^m s_{12})},$$

$$\alpha_{E,31} = \frac{E_3}{H_1} = \frac{-kv(1-v)({}^m q_{11} + {}^m q_{21})^p d_{31}}{{}^p\varepsilon_{33}({}^m s_{12} + {}^m s_{11})kv + {}^p\varepsilon_{33}({}^p s_{11} + {}^p s_{12})(1-v) - 2k^p d_{31}^2(1-v)}. \tag{2.8}$$

Finally, we consider a bilayer laminate that is poled with an electric field E in the plane of the sample. We suppose that the in-plane fields H and δH are parallel, and that the induced electric field δE is measured in the same direction (i.e., along the c-axis). Expression for the α, can be obtained in the following form.

$$\begin{aligned}\alpha_{E,11} = {} & \{[{}^m q_{11}({}^p s_{33}{}^p d_{11} - {}^p s_{12}{}^p d_{12}) + {}^m q_{12}({}^p s_{11}{}^p d_{12} - {}^p s_{12}{}^p d_{11})](1-v) \\ & + [{}^m q_{11}({}^m s_{11}{}^p d_{11} - {}^m s_{12}{}^p d_{12}) + {}^m q_{12}({}^m s_{11}{}^p d_{12} - {}^m s_{12}{}^p d_{11})]vk\}vk(1-v)/ \\ & \{[(1-p)^m\varepsilon_{11} + v^p\varepsilon_{11}][(1-v)^2({}^p s_{11}{}^p s_{33} - {}^p s_{12}^2) \\ & + (1-v)vk({}^m s_{11}{}^p s_{11} + {}^p s_{33}{}^m s_{11} - 2^p s_{12}{}^m s_{12})] \\ & + k^2v^2({}^m s_{11}^2 - {}^m s_{12}^2) - kv(1-v)^2(2^p s_{12}{}^p d_{11}{}^p d_{12} - {}^p s_{33}{}^p d_{11}^2 \\ & - {}^p s_{11}{}^p d_{12}^2 + k^2v^2(1-v)({}^m s_{11}{}^p d_{12}^2 + {}^m s_{11}{}^p d_{11}^2 - 2^m s_{12}{}^p d_{11}{}^p d_{11})\}.\end{aligned} \tag{2.9}$$

Amongst all the cases presented so far, the in-plane ME coefficient is expected to be the largest. This is due to availability of magnetostrictive and piezoelectric phases with high q- and d-values, respectively; and, to the absence of demagnetization fields. We will further use these outcomes later in the estimation of ME parameters for some specific examples.

2.2 Bilayer Structure

Theoretical modeling of low frequency ME effect described above is based on the homogeneous longitudinal strain approach. However, configurational asymmetry of a bilayer implies bending the sample in applied magnetic or electric field and variation in ME response. One of the principal objective of present section is modeling of the ME interaction in a magnetostrictive-piezoelectric bilayer taking into account the flexural strains (Petrov et al. 2009). We calculated ME voltage coefficients α_E for transverse field orientations to provide minimum demagnetizing fields and maximum α_E.

The thickness of the plate is assumed small compared to remaining dimensions. We assume the longitudinal axial strains of each layer to be linear functions of the vertical coordinate z_i to take into account bending the sample. To preserve force equilibrium, the axial forces in the three layers add up to zero, that is,

$$F_{p1} + F_{m1} = 0 \tag{2.10}$$

where $F_{i1} = \int_{-^it/2}^{^it/2} {}^iT_1 \, dz_1$.

The moment equilibrium condition has the form:

$$F_{m1} h_m = M_{m1} + M_{p1}, \tag{2.11}$$

where $M_{i1} = \int_{-^it/2}^{^it/2} z_i^i T_1 \, dz_1$.

Simultaneous solving (2.8) and (2.10) enables finding the axial stress components in the piezoelectric layer ${}^pT_1$ and ${}^pT_2$. Then the expression for ME voltage coefficient can be expressed using the open circuit condition

$$\alpha_{E,31} = \frac{E_3}{H_1} = -\frac{{}^pd_{31} \int_{-^pt/2}^{^pt/2} ({}^pT_1 + {}^pT)_2 \, dz}{t \, H_1 {}^p\varepsilon_{33}},$$

where $t = {}^mt + {}^pt + {}^st$ is the total thickness of considered structure.

Using the 1-D approximations of (2.1)–(2.10) enables one to obtain an explicit form of expression for ME voltage coefficient

$$\alpha_{E,31} = \frac{\left[\left(1 - {}^{p}K_{31}^{2}\right){}^{p}s_{11} + {}^{m}s_{11}{}^{p}t^{3}/{}^{m}t^{3}\right]{}^{m}q_{11}{}^{p}d_{11}/{}^{p}\varepsilon_{33}}{{}^{p}s_{11}\left(1 - {}^{p}K_{31}^{2}\right)\left[2{}^{p}t/{}^{m}t\,{}^{m}s_{11}\left(2 + 3{}^{p}t/{}^{m}t + 2{}^{p}t^{2}/{}^{m}t^{2}\right) + \left(1 - {}^{p}K_{31}^{2}\right){}^{p}s_{11}\right] + {}^{m}s_{11}^{2}{}^{p}t^{4}/{}^{m}t^{4}} \quad (2.12)$$

In case of neglecting the flexural strains, it is easily shown that expression for ME voltage coefficient reduces to well-known expression of Sect. 2.1 which was obtained with the assumption of homogeneous longitudinal strains.

2.3 Examples of Multilayer Structures

The preceding comprehensive theoretical treatment resulted in expressions of the ME voltage coefficients for three different orientations of fields, which were the ones of most importance, including: longitudinal, transverse, and in-plane longitudinal. The most significant features of the model are as follows: (i) Consideration of three different field configurations. This allows for the determination of a single-valued interface parameter k, facilitating quantitative characterization of the bilayer interface. (ii) Consideration of a new field configuration, i.e., in-plane longitudinal fields that has very strong ME coupling. And, (iii) consideration of the effect of a finite magnetic permeability on the magnetostriction of the magnetic subsystem: which was ignored in prior investigations.

Next, we apply the theory for the calculation of ME coupling in layered composites. Consider the materials couple cobalt ferrite and lead zirconate titanate (CFO–PZT), which is a system that has been of significant prior interests. Since the value of α_E depends notably on the concentration of the two phases, the ME voltage coefficient has been determined as a function of the volume fraction v of the piezoelectric phase in composite. Material parameters used for estimates are given in Table 2.1. Results of calculations using the model are illustrated in Fig. 2.2, which were obtained by assuming an ideal interface coupling ($k = 1$).

Results of $\alpha_{E,31}$ versus PZT volume fraction reveals a double maximum that is due to fact that the strain produced by the ferrite consists of two components: longitudinal and flexural. For a symmetric structure such as trilayer, there are no flexural strains and the maximum ME coefficient occurs for $V = 0.6$ (Bichurin et al. 2003). Since the flexural strain is opposite in sign compared to longitudinal one and reaches its maximum value for $V = 0.6$, the two types of strains combine to produce suppression of $\alpha_{E,31}$ at $V = 0.6$ and a double maximum in the ME coefficient as in Fig. 2.2. In what follows, we consider theoretical models of low frequency ME coupling for symmetric structures taking into account that such structures result in higher values of ME coefficients.

The variation of $\alpha_{E,33}$ with v for various values of coupling parameter k is shown in Fig. 2.3a for a symmetric structure which excludes the flexural deformations. The magnitude of $\alpha_{E,33}$ decreases with decreasing k, and $v_{\max}$ shifts to

Table 2.1 Material parameters (compliance coefficient s, piezomagnetic coupling q, piezoelectric coefficient d, and petmittivity ε) for lead zirconate titanate (PZT), cobalt ferrite (CFO), and lanthanum strontium manganite used for theoretical values

Material	s_{11} (10^{-12} m^2/N)	s_{12} (10^{-12} m^2/N)	s_{13} (10^{-12} m^2/N)	s_{33} (10^{-12} m^2/N)	q_{33} (10^{-12} m/A)	q_{31} (10^{-12} m/A)	d_{31} (10^{-12} m/V)	d_{33} (10^{-12} m/V)	$\varepsilon_{33}/\varepsilon_0$
PZT	15.3	−5	−7.22	17.3	–	–	−175	400	1,750
CFO	6.5	−2.4			−1,880	556	–	–	10
LSMO	15	−5			250	−120	–	–	10

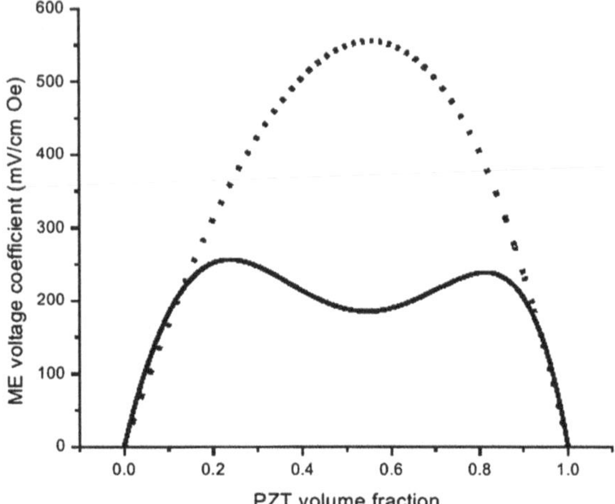

Fig. 2.2 PZT volume fraction dependence of transverse ME voltage coefficient $\alpha_{E,31} = \delta E_3/\delta H_1$ for a perfectly bonded ($k = 1$) bilayer (*solid line*) and symmetric structure (*dot line*) of CFO and PZT

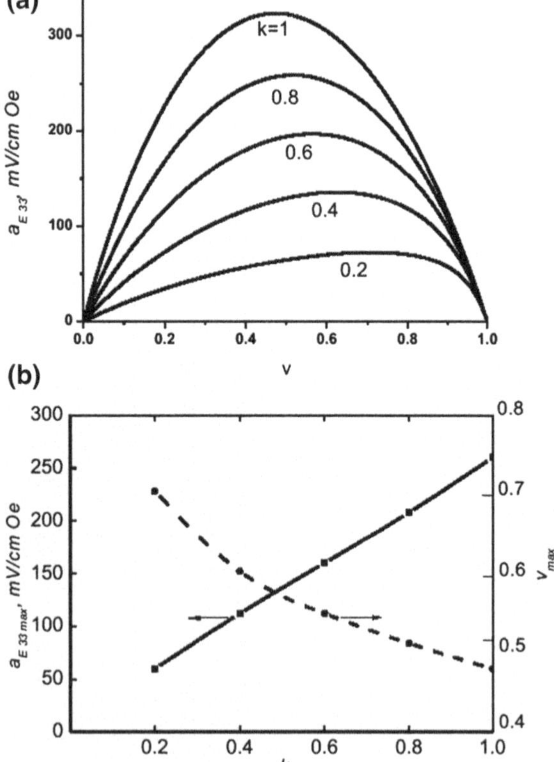

Fig. 2.3 **a** Estimated dependence of longitudinal ME voltage coefficient on interface coupling k and volume fraction v for symmetric structure of CFO and PZT. **b** Variation with k of maximum $\alpha_{E,33}$ and the corresponding v_{max}

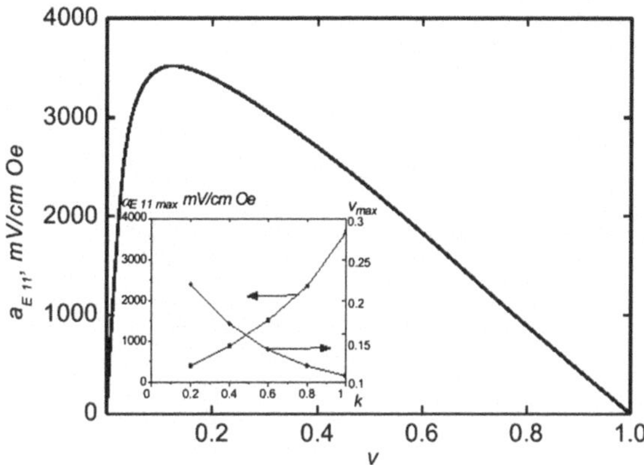

Fig. 2.4 ME voltage coefficient for a perfectly bonded (k = 1) symmetric structure of CFO and PZT for in-plane longitudinal field orientation. The poling field and dc and ac magnetic and electric fields are in the sample plane and parallel to each other. *Inset* shows variation of maximum $\alpha_{E,33}$ and the corresponding v_{max} with k

PZT-rich compositions. Figure 2.3b shows the dependence of the maximum value in $\alpha_{E,33}$ on k, where calculations are illustrated for various values of v_{max}. With increasing k, a near-linear increase was found in the maximum value of $\alpha_{E,33}$. For transverse fields, the maximum α_E is 40 % higher than that of $\alpha_{E,33}$. This is due to the strong parallel piezomagnetic coupling q_{11} which determines α_E, relative to that of q_{31} which determines $\alpha_{E,33}$.

Next, we consider the ME effect in CFO–PZT for the in-plane longitudinal field orientation.

The most significant prediction of the present model is that the strongest ME coupling should occur for in-plane longitudinal fields, as shown in Fig. 2.4. One can easily see in Fig. 2.4 that when the field is switched from longitudinal to in-plane longitudinal that the maximum value of the relevant ME coefficient increases by nearly an order of magnitude: $\alpha_{E,max} = 325$ mV/cm Oe for the longitudinal orientation, whereas $\alpha_{E,11} = 3{,}600$ mV/cm Oe for the in-plane longitudinal. The v-dependence of $\alpha_{E,\,11}$ reveals a rapid increase in the ME coefficient to a maximum value of $\alpha_{E,11} = 3{,}600$ mV/cm Oe for $v = 0.11$, which is followed by a near-linear decrease with further increase of v. Such an enhancement in the in-plane longitudinal coefficient relative to the longitudinal one is understandable due to (i) the absence of demagnetizing fields in the in-plane configuration, and (ii) increased piezoelectric and piezomagnetic coupling coefficients compared to longitudinal fields. The down-shift in the value of v_{max} (from 0.5 to 0.6 for longitudinal and transverse fields to a much smaller value of 0.1) is due to the concentration dependence of the effective permittivity.

Another layered structure of importance is nickel ferrite (NFO)–PZT. Although NFO is a soft ferrite with a much smaller anisotropy and magnetostriction than CFO, efficient magneto-mechanical coupling in NFO–PZT gives rise to ME voltage coefficients comparable to those of CFO–PZT. Using the model presented in this chapter, we can estimate α_E for NFO–PZT for different field orientations and conditions, similar to that for CFO–PZT.

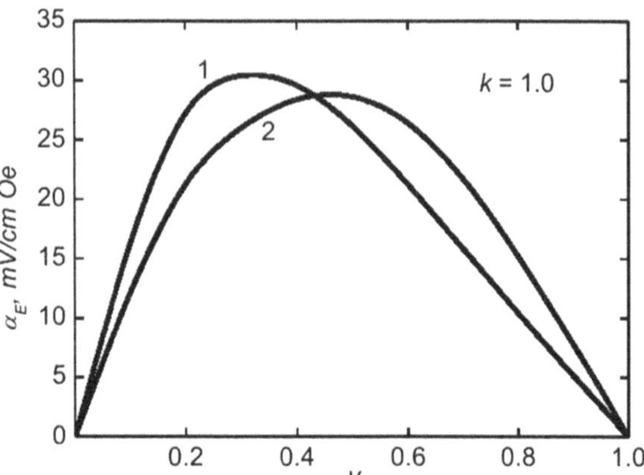

Fig. 2.5 *1* Longitudinal and *2* transverse ME voltage coefficients as functions of PZT volume fraction for symmetric layered structure of $La_{0.3}Sr_{0.7}MnO_3$ (LSMO) and PZT for interface coupling parameter k =1

Finally, we consider composites that have lanthanum strontium manganites for the magnetostrictive phase. Lanthanum manganites with divalent substitutions have attracted considerable interest in recent years due to double exchange mediated ferromagnetism, metallic conductivity, and giant magnetoresistance. The manganites are potential candidates for ME composites because of (i) high magnetostriction and (ii) metallic conductivity that eliminates the need for a foreign electrode at the interface. Figure 2.5 shows the longitudinal and transverse ME voltage coefficients for unclamped $La_{0.3}Sr_{0.7}MnO_3$ (LSMO)–PZT laminate that assumes ideal coupling at the interface and no bending strain. In this case, the values of the ME coefficients are quite small compared to that of ferrite–PZT: this is due to weak piezomagnetic coefficients and compliances parameters for LSMO. The ME coefficient for in-plane longitudinal fields and the effects of clamping for different field orientations were similar in nature to those for ferrite–PZT layered structure, and thus are not discussed in any detail here.

It is important to compare the theoretical predictions, illustrated above, with experimental data. Let us consider first a bilayer of CFO–PZT taking into account the flexural deformations. Figure 2.6 shows α_E as a function of v. These data were obtained at low frequencies (100–1,000 Hz). The desired volume fractions v was achieved by careful control of the layer thickness. Data show an increase in α_E with v until a maximum is reached. However, these data clearly demonstrated that the actual experimental value is an order of magnitude smaller than that predicted in Figs. 2.2 and 2.3 (assuming $k = 1$). It is, therefore, reasonable to compare the data with calculated values of α_E as a function of v using a reduced interface coupling parameters of $k = 0.1$: in this case, agreement between theory and experiment can be seen, as shown in Fig. 2.6. The key inference that can be made concerns the inherently poor interface coupling for CFO–PZT, irrespective of sample synthesis techniques. We address possible causes for this poor coupling later in this section.

A third materials couples, LSMO–PZT, is considered in Fig. 2.7, which shows α_E as a function of v for longitudinal and transverse fields. The α_E values are the

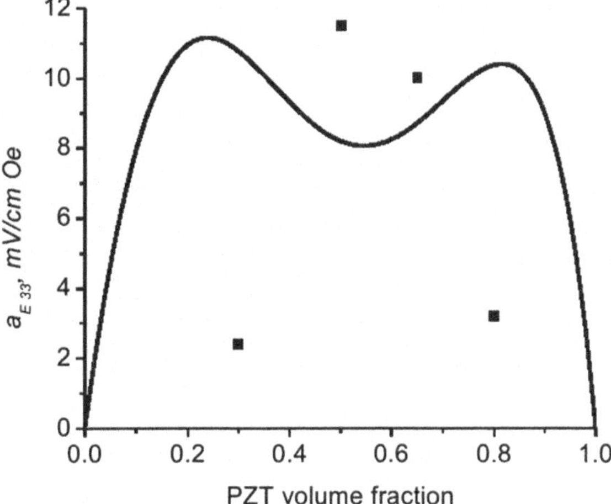

Fig. 2.6 PZT volume fraction dependence of longitudinal ME voltage coefficient of CFO–PZT bilayer. Solid line are theory for $k = 0.1$ and points are experiment

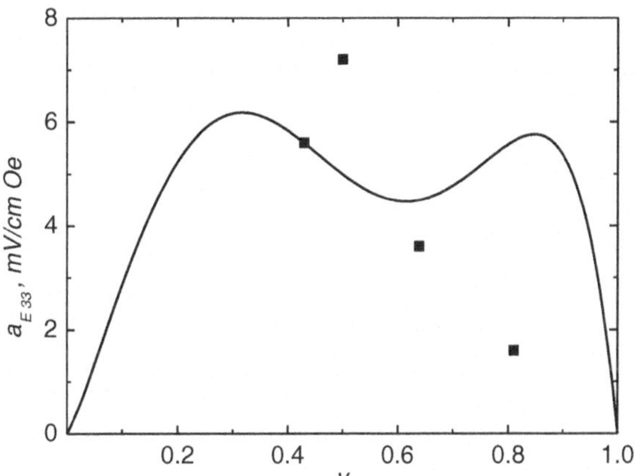

Fig. 2.7 PZT volume fraction dependence of transverse **a** and longitudinal **b** ME voltage coefficients for LSMO–PZT bilayer: *solid lines* are the theory for $k = 0.2$ and points are experiment

smallest amongst the three systems considered here. Calculated values assuming $k = 1$ were found to be quite high compared to the data, rather it was found that nonideal values of $k = 0.2$ gave reasonable agreement with the data. Thus, one can readily infer poor interfacial coupling in LSMO–PZT, similar to that for CFO–PZT.

Finally, we should comment on a possible cause of poor interfacial coupling for CFO–PZT and LSMO–PZT, and ideal coupling for NFO–PZT. The parameter k can be expected to be sensitive to mechanical, structural, chemical, and electromagnetic parameters at the interface. We attribute unfavorable interface conditions in CFO–PZT and LSMO–PZT to inefficient magneto-mechanical coupling. The magneto-mechanical coupling k_m is given by $k_m = (4\pi\lambda'\mu_r/E)^{1/2}$; where λ' is the dynamic magnetostrictive constant and μ_r is the reversible permeability, and E is Young's modulus. In ferrites, under the influence of a dc magnetic bias H and ac magnetic field δH, domain wall motion and domain rotation contribute to the Joule magnetostriction and consequently to the effective linear piezomagnetic coupling. A key requirement for strong coupling is unimpeded domain wall motion and domain rotation. A soft ferrite with a high initial permeability (i.e., low anisotropy), such as NFO, will have key materials parameters favoring a high k_m, and consequently, strong ME effects. Measurements have shown that NFO has an initial permeability of 20, whereas that of LSMO and CFO is 2–3. Thus, one can infer a plausible simple explanation of the near interfacial parameter for NFO–PZT is (in part) favorable domain motion.

In deriving the above expression, we assumed the electric field to be zero in magnetic phase since magnetostrictive materials that are used in the case under study have a small resistance compared to piezoelectric phase. Estimate of ME voltage coefficient for CFO–PZT layered structure gives $\alpha_{E,\,33} = 325$ mV/cm Oe providing that the bending strains are ignored. However, considering CFO as a dielectric results in $\alpha_{E,\,33} = 140$ mV/cm Oe (Osaretin and Rojas 2010) while the experimental value doesn't exceed 74 mV/cm Oe (Harshe et al. 1993). We believe CFO should be considered as a conducting medium compared to dielectric PZT in the low-frequency region in accordance with our model. The discrepancy between theoretical estimates and data can be accounted for by features of piezomagnetic coupling in CFO and interface coupling of bilayer (Bichurin et al. 2003).

2.4 Bulk Composites

Design of new ME composites assumes the use of reliable theoretical models, allowing prediction of properties for various materials couples and over a range of laminate parameters. Manufacturing methods of all-ceramic composites are based on an initial mixing of starting powders batched in proportion to the composite volume fraction, followed by pressing and densification/sintering to a net-shape. Clearly, if the concentration of one of the constituent phases is small, then that phase will consist of isolated particles in a matrix. Following accepted classification nomenclature (Newnham et al. 1978) this composite should be referred to as a 0–3 type, as one phase is isolated (i.e., connected in zero dimensions) and the second is interconnected in three dimensions. If the volume fraction of the secondary phase in the matrix is increased, and a percolation limit is reached, then it is classified as a 1–3 type composite. If the secondary phase then crosses that initial percolation limit, and subsequently begins to be interconnected in two dimensions, the composite connectivity is known as the 2–3 type. We mention these things at this time to make the point that the same ceramic manufacturing technology allow the fabrication of a wide range of relative volume fractions of the different phases in an all-ceramic composite, and consequently to various possible types of dimensional interconnectivities. Accordingly, it is very important to choose the correct method of calculation for effective constants of a composite at various relative volume fractions of components.

Unfortunately, exact solutions of three-dimensional problems related to the calculation of effective constants of inhomogeneous systems are unknown. Therefore, there is presently no precise structural classification of composites. Within the limited theory of heterogeneous systems of two-phase composites, there are two principle approaches to approximate solutions: matrix systems and two-component mixtures, for which behavior of effective parameters depending on concentration continuously.

In the case of matrix systems, modification of the concentration from 0 to 1 does not change the qualitatively structure of the composite: at any concentration, one of the components must form a coherent matrix that contains isolated particles of the second component. The system always remains essentially non-central, and matching formulas for an evaluation of effective constants give their continuous dependence on concentration in the entire range from 0 to 1. We should note that the application of these formulas to the calculation of effective constants of composites is not always justified.

The case of two-component mixtures is characterized by a qualitative modification of the structure of the composite, as the concentration is changed. Such systems are characterized, as is well known, by critical concentrations at which point there are important property changes such as metal–insulator or rigid-plastic transformations. The metal–insulator transformation occurs in a composite consisting of an insulating and conductive phases. Assume that the insulating phase is initially the matrix and that the conductive one consists of isolated particles. In this case, initially the composite is insulating; however, when the percolation limit is crossed, the conducting particles form an interconnected conduction pathway, dramatically lowering the resistivity near a critical volume fraction. In the second example (rigid-plastic transformation), it is supposed that the composite is a mixture in which the elastic compliance of one of the constituent phases tends to infinity (for example, a porous composite). This composite type posses a critical concentration of the second phase, above which the rigid framework of the composite loses its stability. It should be straightforward to see that any bulk composites will have numerous effective materials properties, all of which change with relative phase volume fraction from in a manner independent of other properties.

As an example, we consider a composite with a 3–0 type connectivity. Cubic models for ferrite–ferroelectric composites with a connectivity of 3–0 and 0–3 have been considered by Harshe et al. (1993). Numerically, the ME coefficient is equal to the ratio of the electric field induced on the composite by an applied magnetic field: the ME coefficient is equal to E_3/H_3. It is necessary to realize that the magnetic field was applied only to the ferrite phase: i.e., $E_3/{}^{m}H_3$ where ${}^{m}H_3$ is the local magnetic field on the ferrite phases which may exceed that applied to the entire composite. Harshe's study only considered the case of free cubic cells, and effective parameters of the composite in known model systems were not determined. However, in real composites, we must consider the case of non-free cells. It is also very important to use any such model to predict the effective composite parameters. In the following section, we present a generalized model for ferrite–piezoelectric composites that allows one to define and predict the effective parameters of said composite using given conditions.

The properties of this ME composite will depend on the parameters of the corpuscles, and also on the terminal conditions. Now, let us suppose that the geometrical model for ME composites in this figure is miniaturized to fine scales. If the given cubic model ME composite is considered as a material consisting of consecutive and parallel connections of cubic cells with legs of unit length, then it is obvious by the definition of properties of a composite that it is possible to consider only one cubic cell rather than the entire ensemble of cells. The magnetostriction phase is enclosed by piezoelectric ones along different directions.

$$\begin{aligned} S_1 &= -s_{c1}s_{11}T_1, \\ S_2 &= -s_{c2}s_{22}T_2, \\ S_3 &= -s_{c3}s_{33}T_3, \end{aligned} \tag{2.13}$$

where $s_{ci} = {}^{c}s_{ii}/s_{ii}$ is the relative compliance of surroundings, and $c_{s_{ii}}$ is effective compliance of composite.

PZT volume fraction dependence of effective ME susceptibility is shown in Fig. 2.8 (Petrov et al. 2004, 2007).

The dependence of the effective ME voltage coefficient, defined as $\alpha_{E,33} = -\alpha_{33}/\varepsilon_{33}$, on the piezoelectric phase volume fraction can then be easily obtained, as shown in Fig. 2.9. These graphical solutions then allow one to determine the piezoelectric and magnetostrictive phase volume fractions that yield maximum values for the effective ME susceptibility (Fig. 2.8) and the ME voltage coefficient (Fig. 2.9).

The values of the ME voltage coefficient in Fig. 2.10 coincide with previously published data (Harshe et al. 1993), demonstrating the usefulness of the predictions. As follows from Fig. 2.10, the ME voltage coefficient was approximately 20 % greater than that calculated from the experimental data using the model. This is explained by the fact that the internal (local) magnetic field in the ferrite component is considerably different than that of the externally applied magnetic field.

Measurements of the ME voltage coefficient have been performed for bulk composites of NFO–PZT, using the experimental methodology mentioned above. Data are shown in Fig. 2.11 for the ME voltage coefficient as a function of the piezoelectric phase volume fraction. These measured values are much lower in magnitude than the theoretical ones predicted for the free composite condition. However, considering a clamped condition defined by the matching $s_{c11} = s_{c22} = s_{c33} = 0.3s_{33}$, agreement between theory and experiment was found, as illustrated in Fig. 2.11. These results indicate in real 0–3 ferrite–piezoelectric ceramic composites mixtures that the component phase grains are mechanically clamped by neighboring grains and by environmental boundary conditions.

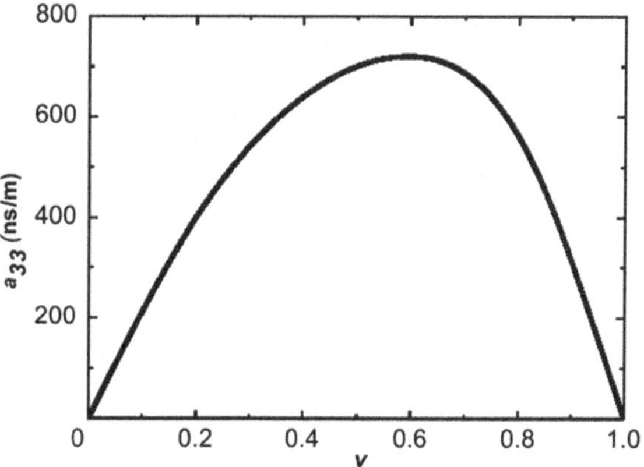

Fig. 2.8 PZT volume fraction dependence of ME susceptibility for CFO–PZT composite with connectivity 3–0

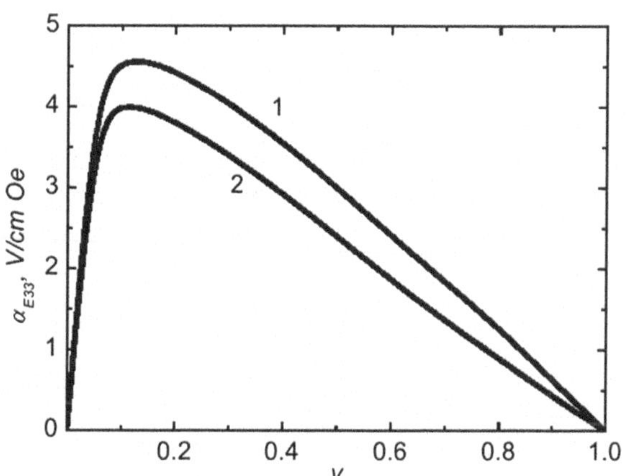

Fig. 2.9 PZT volume fraction dependence of ME voltage coefficient for CFO–PZT composite with connectivity 3–0 according to model (Petrov et al. 2004) *1* and model (Harshe et al. 1993) *2* for $\alpha_{E33} = E_3/^m H_3$ and material parameters from Table 2.1

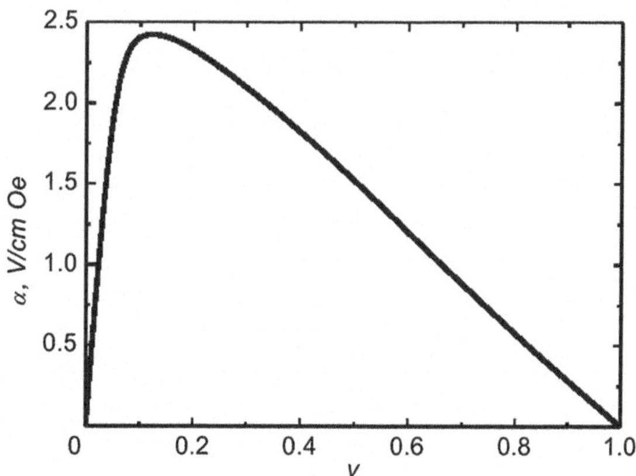

Fig. 2.10 Concentration dependence of $\alpha_{E33} = E_3/^m H_3$ for composite with material parameters from (Petrov et al. 2004)

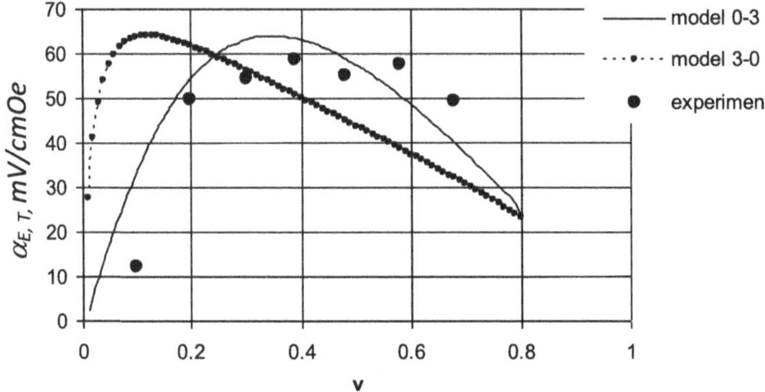

Fig. 2.11 ME effect in bulk composite of NFO and PZT

2.5 Magnetoelectric Effects in Compositionally Graded Layered Structures

A key drawback of functional devices based on ME materials is the need of prior polarization of ferroelectric phase and magnetic bias field of magnetic phase. Poling consists in heating the sample to Curie temperature and slow cooling in a static electric field. The bias magnetic field magnitude should correspond to maximal piezomagnetic coupling coefficient which provides the maximal ME coefficient. Thus supplementary constructional elements are necessary to provide poling and biasing the sample.

Here we prepare a composite based on magnetization-graded magnetic phase from compositionally graded ferromagnetic and polarization-graded piezoelectric phase from compositionally graded ferroelectric. Using the polarization-graded piezoelectric phase is known to result in a built-in static electric field. The amplitude of this field is determined by polarization gradient, $E_i = -\frac{1}{\varepsilon_0}\int\limits_0^L \frac{\partial P}{\partial x}\,dx$ where L is the sample length. Analogously, magnetization gradient gives rise to a built-in static magnetic field, $H_i = -\frac{1}{\mu_0}\int\limits_0^L \frac{\partial M}{\partial x}\,dx$. The above expressions show that the proper choice of polarization and magnetization gradients and sample length enables one to get the built-in electric and magnetic fields needed for obtaining the maximal piezoelectric and piezomagnetic coupling coefficients. It should be noted that there is no need in preliminary poling and magnetic biasing the sample (Petrov and Srinivasan 2008; Mandal et al. 2011).

As an example, we consider a magnetostrictive-piezoelectric bilayer of nickel–zinc ferrite and PZT. Bilayer includes the magnetic layer which is compositionally graded along the sample plane and the piezoelectric layer which is compositionally graded perpendicular to the sample plane. The output voltage induced by ME

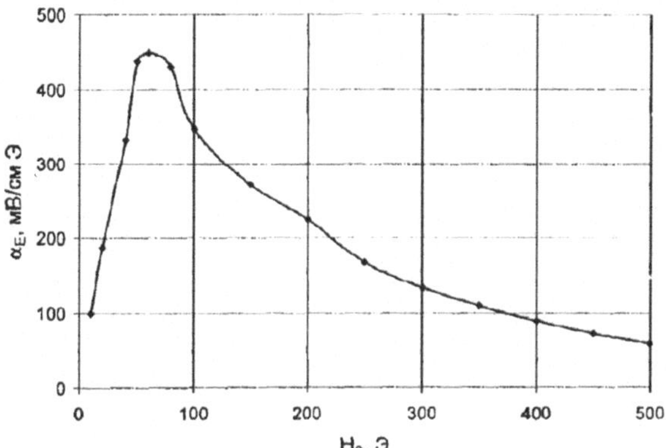

Fig. 2.12 Bias field dependence of ME voltage coefficient for bilayer of compositionally graded nickel–zink ferrite and PZT

coupling is measured across the sample thickness. Estimated bias field dependence of ME voltage coefficient is shown in Fig. 2.12 for equal volume fractions of ferrite and PZT.

Variation of zinc content from 0.3 to 0.5 gives rise to a magnetization gradient in the sample plane that results in a built-in magnetic field of 44 Oe. Estimates show that variation of zinc content from 0.3 to 0.5 enables one to increase the built-in magnetic field up to 60 Oe. Figure 2.12 shows that bias field of 60 Oe provides obtaining ME voltage coefficient of 450 mV/cm Oe.

2.6 Magnetoelectric Effect in Dimensionally Graded Laminate Composites

Dimensionally gradient piezoelectric plate with thickness of 1 mm was fabricated by mechanical polishing and dicing technique, as shown in Fig. 2.13a (Park et al. 2012). Piezoelectric plates with composition $Pb(Zn_{1/3}Nb_{2/3})_{0.2}(Zr_{0.5}Ti_{0.5})_{0.8}O_3$ [PZNT] were synthesized by conventional mixed oxide method. Piezoelectric constant of poled PZNT plates was found to be 500 pC/N and the dielectric constant was 2,219 at 1 kHz. The piezoelectric voltage constant (g_{33}) was of the order of 23.41×10^{-3} Vm/N. On this PZNT plate, 25 μm-thick Metglas (2605SA1, Metglas Inc, USA) sheets of desired dimensions were attached using epoxy (West System, USA) with the curing temperature of 80 °C. Impedance spectrum of the composites was measured by LCR meter (HP 4194A). For ME voltage coefficient measurement in longitudinal–transversal (L–T) mode configuration, an electromagnet was used to apply the DC magnetic field and the samples were placed in the center of the Helmholtz coil under an AC magnetic field (H_{ac}). The voltage induced on the laminate was monitored by using a lock-in amplifier. The magnetostriction was evaluated by using the strain gauge and Wheatstone bridge.

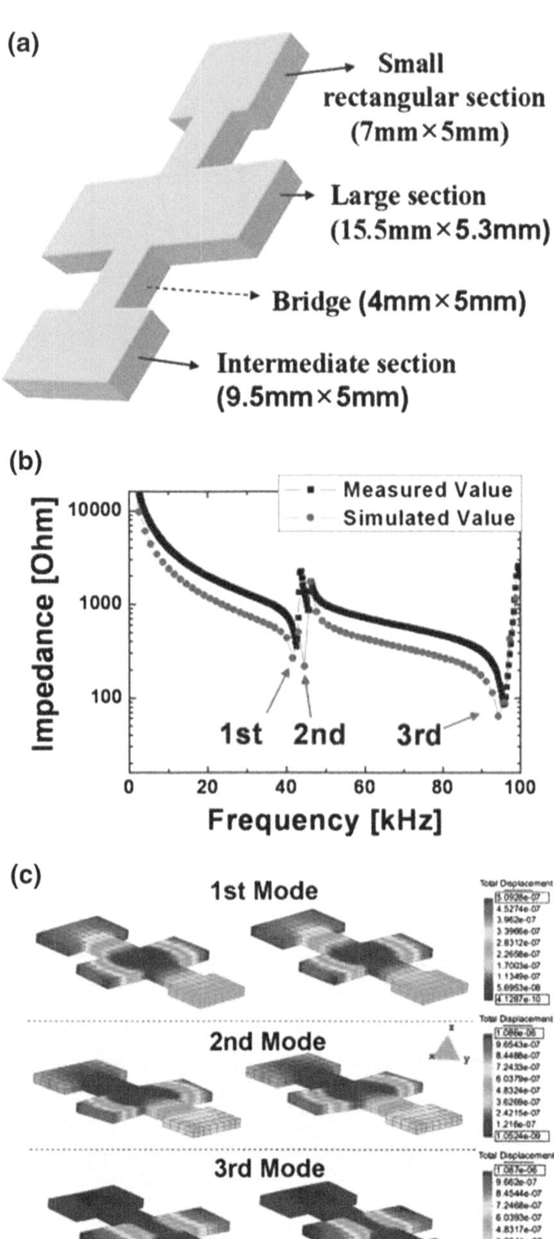

Fig. 2.13 **a** Schematic diagram of asymmetric piezoelectric plate, **b** impedance spectrums from simulation and measurement, and **c** resonance displacements at 42, 44, and 94.3 kHz

Impedance spectrum of asymmetric piezoelectric plate was measured to identify the EMR range. The first, second, and third resonances were found to be 42, 44, and 94.3 kHz, respectively. In comparison, measured impedance spectrums for the asymmetric piezoelectric plate exhibited resonance peaks at 42, 44, and 96 kHz, as shown in Fig. 2.13b. Thus, the results between simulation and measurement were in good agreement. The first mode at 42 kHz was associated with biaxial bending of the large rectangular section and the intermediate rectangular section, while the second mode at 44 kHz was related to the biaxial bending of the large rectangular section and the small rectangular section. The third mode at 99 kHz came from the lateral displacement of the large rectangular section, as shown in Fig. 2.13c.

Figure 2.14a shows the fabricated ME laminate composite based on Fig. 2.13a. On top of the PZNT layer, four layers of Metglas with the area of 13×7 mm^2 were attached at the middle rectangular section (Section A), 30 layers of Metglas with area of 20×7 mm^2 were attached at the larger rectangular section (Section B), and five layers of Metglas with area of 7×7 mm^2 were attached at the intermediate rectangular section (Section C), as shown in Fig. 2.14a. There are two variables which could be adjusted to achieve a wideband ME response. First, if the rectangular area of two sections in piezoelectric plate is different than the one with smaller number of Metglas layers will show higher ME coefficient. Second, if the numbers of Metglas layers are same, the rectangular section with the smaller area will show smaller ME coefficient. Thus, by adjusting the ratio of Metglas layers on various rectangular sections, a composite ME response with flat behavior can be obtained.

Figure 2.14b shows the magnetostriction (S_{ij}) and piezomagnetic (q_{ij}) coefficient for varying dimensions and stack configurations of Metglas. In this figure, S_{11} corresponds to longitudinal in-plane magnetostriction parallel to H_{DC} and q_{11} is the longitudinal in-plain piezomagnetic coefficient corresponding to the differential of S_{11}. The maximum in-plane magnetostriction was found to be 28 ppm regardless of Metglas stack configurations and dimensions; however, the strain behavior was strongly dependent on the stack configuration and dimensions. Four layers of Metglas with the area of 13×7 mm^2, 30 layers of Metglas with area of 20×7 mm^2, and five layers of Metglas with area of 7×7 mm^2 showed piezomagnetic coefficient corresponding to 0.38, 0.21, and 0.16 ppm/Oe at 70, 150, and 204 Oe of H_{DC}, respectively. These piezomagnetic behaviors will result in strong elastic coupling with the piezoelectric sections.

Figure 2.14c shows the measured ME response from the composite structure as a function of magnetic DC bias under the condition of $H_{AC} = 1$ Oe at $f = 1$ kHz. The peak at 70 Oe was associated with section A. The second peak of the ME coefficient at 150 Oe was associated with section C. The third peak of the ME coefficient at 209 Oe was associated with section B. The value of DC bias at the peak in piezomagnetic coefficient corresponds to that for the peak in ME coefficient. Further, it was found that not only piezomagnetic coefficient but also piezoelectric dimensions affected elastic coupling between the Metglas and piezoelectric sections compared to Fig. 2.14b and c. Furthermore, if only section A and section B were combined, there will be a valley in the intermediate range. By inserting section C, the formation of flat ME band was facilitated. The overall ME

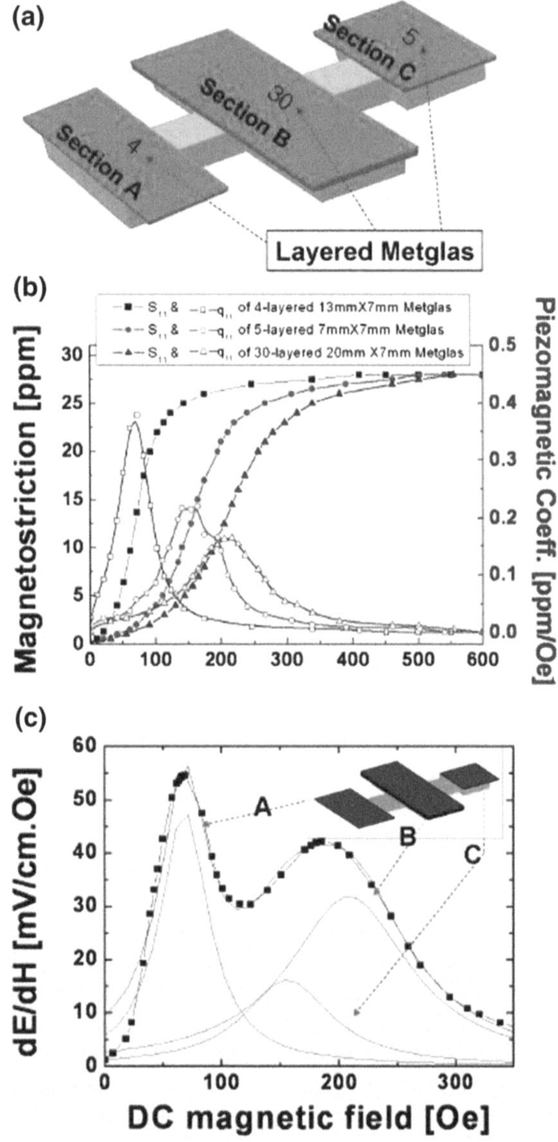

Fig. 2.14 **a** Schematic diagram of laminated ME composite, **b** magnetostriction (S_{11}) and piezomagnetic coefficient (q_{11}) for varying dimensions and stack configurations of Metglas, **c** ME response as a function of DC magnetic field under the constant condition of $H_{ac} = 1$ Oe at $f = 1$ kHz

response as a function of magnetic DC bias was composite of three individual responses.

The impedance and phase angle spectrums for this laminate are shown in Fig. 2.15a. After Metglas was attached on the PZNT plate, multiple resonances occurred at 14, 28, 41, 44, and 107 kHz as marked with arrows in Fig. 2.15a. Resonant modes at 14 and 28 kHz were found to be related to the bending vibrations of all the sections. The middle section dominated the vibration mode at

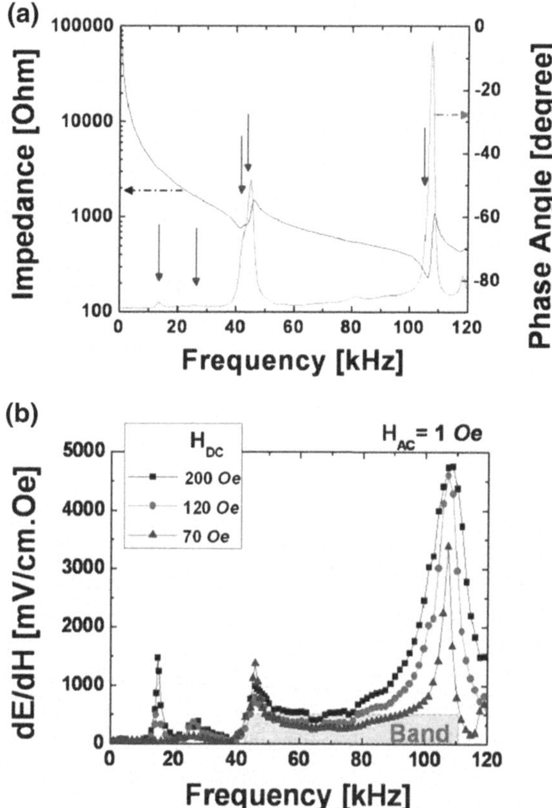

Fig. 2.15 **a** Impedance and phase angle spectrums of the ME composite, **b** ME response as a function of frequency

14 kHz while the small rectangular section dominated the vibration mode at 28 kHz. The resonance at 41 kHz was found to be combination of the bending motions of large rectangular section (section B) and middle rectangular section (section A). The resonance at 44 kHz was related to combined bending response from the bridge and small rectangular section (section C). The resonant modes of ME composite at 41 and 44 kHz were correlated to the first and second resonance modes of PZNT layer as determined in Fig. 2.13b. The resonant mode at 107 kHz had a similar vibration mode as that at 44 kHz.

The ME output voltage from composite was measured as a function of frequency under constant $H_{AC} = 1$ Oe but varying $H_{DC} = 70$, 120, and 200 Oe, as shown in Fig. 2.15b. Interestingly, ME response as a function of frequency had similar behavior as that of phase angle shift shown in Fig. 2.15a and b. The ME peaks are shown at 14, 28, 45, and 107 kHz corresponding to the resonance frequencies observed in the impedance measurement. The peak ME response at 45 kHz was associated with resonances at 41 and 44 kHz. The maximum ME coefficient was found to be 4,740 mV/cm Oe at $f = 107$ kHz under $H_{DC} = 200$ Oe. Interestingly, the bands were successfully formed in the range of

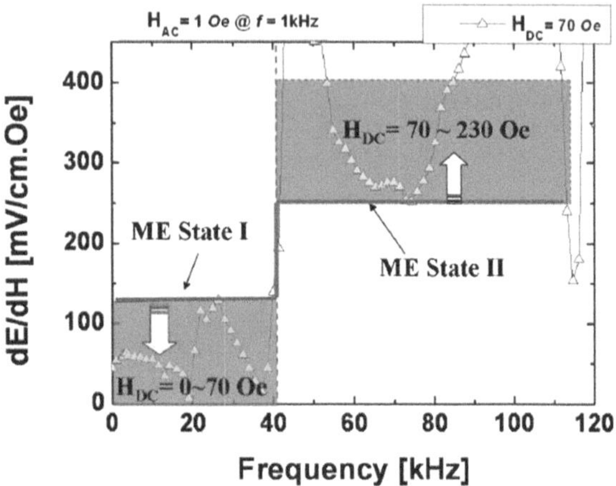

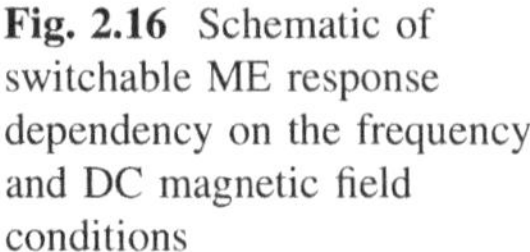
Fig. 2.16 Schematic of switchable ME response dependency on the frequency and DC magnetic field conditions

41–110 kHz. The composite showed high ME coefficient of 260 mV/cm Oe after the resonance peak of 41 kHz, regardless of the applied DC magnetic field. These widely extended bands were derived from the combination of the resonances at 41, 44, and 107 kHz.

The maximum ME coefficients of 1,400 and 4,740 mV/cm Oe was found at 45 and 107 kHz respectively. The band at 107 kHz exhibited ME coefficient higher than 3,000 mV/cm Oe from 52 to 242 Oe of H_{DC} while the band at 45 kHz exhibited ME coefficient higher than 780 mV/cm Oe from 40 to 230 Oe of H_{DC}. The bands were obtained regardless of applied DC and AC magnetic fields. The wideband was formed in both conditions of the frequency range of 41–110 kHz and DC magnetic ranges of 40–230 Oe and 52–242 Oe at $f = 45$ and 107 kHz, as shown in Figs. 2.15b.

Interestingly, there are two noticeable states in the ME frequency response in Fig. 2.15b. Under the constant DC magnetic field condition of 70 Oe, the composite showed flat ME responses in both frequency ranges of 1–11 kHz (state I) and 41–110 kHz (state II) as shown in Fig. 2.16. Figure 2.16 shows the schematic diagram of resulting ME response as a function of frequency and the applied DC magnetic field conditions. State I shows a band lower than 129 mV/cm Oe before $f = 41$ kHz while State II shows a band higher than 260 mV/cm Oe after $f = 41$ kHz under the constant condition of $H_{DC} = 70$ Oe. The ME coefficient of State II was 2× higher than that of State I. Consequently, ΔME ($= ME_{StateII} - ME_{StateI}$) was 131 mV/cm Oe under the constant $H_{DC} = 70$ Oe. These states were found to be adjustable with changing H_{DC} conditions. The maximum value of the ME_{StateI} can be reduced by decreasing the DC magnetic field and ranges between 0 and 129 mV/cm Oe. On the other hand, the minimum value of the $ME_{StateII}$ can be elevated by increasing DC magnetic field condition in the range of 260–406 mV/cm Oe. Thus by tuning the frequency and H_{DC} ΔME in the range of 131 and 406 mV/cm Oe can be achieved which clearly shows the tunability of this device.

These signals are strong enough to allow two distinguishable states. The clear sensing margin was seen between state I and state II: the ME values in state II were at least two times higher than that in state I. This interesting and promising ME behavior can be exploited in several applications. The wideband behavior can be a candidate for magnetic field controlled switches as well as ME harvesters. In the case of the magnetic field-controlled switches, the resonant frequency becomes the cut-off condition and the state I and state II can be considered as "off" and "on," respectively.

We attempt to model the response of the laminate structure by calculating the in-plane strain and stress components for all sections and at the end combining them together to find the overall solution. The total vibration spectrum of the laminate composite consists of several contributions. The first mode is supposed to be associated with simultaneous bending vibrations of the small rectangular section with its bridge and the middle rectangular section. The equation of bending motion of ith-area (1 and 2 areas correspond to bridge and section) has the form:

$$\nabla^2\nabla^2 w_i + \frac{\rho_i\, t_i}{D_i}\frac{\partial^2 w_i}{\partial t^2} = 0, \tag{2.14}$$

where $\nabla^2\nabla^2$ is biharmonic operator, w_i is the displacement in z-direction, t_i is thickness, ρ_i is average density of i-area, and D_i is cylindrical stiffness. The strain's components can be expressed in terms of displacement as $S_{1i} = -z\frac{\partial^2 w_i}{\partial x^2}$ and $S_{2i} = -z\frac{\partial^2 w_i}{\partial y^2}$. The stress components can be expressed in terms of strains as:

$$\begin{aligned}({}^pS_k)_i &= {}^ps_{kj}\left({}^pT_j\right)_i + {}^pd_{31}\,{}^pE_{3i};\\ ({}^mS_k)_i &= {}^ms_{kj}\left({}^mT_j\right)_i + ({}^mg_{k1})_i^m B_{1i};\end{aligned} \tag{2.15}$$

where S_{1i} and T_{1i} are strain and stress components for i-area, E_{3i} is the component of electric field, H_{1i} is the component of magnetic field, s_{kj}, is compliance at constant electric field for piezoelectric and at constant magnetic induction for magnetic component, g_{k1} and d_{31} piezomagnetic and piezoelectric coefficients correspondingly. The superscripts p and m correspond to piezoelectric and piezomagnetic layers. Solving (2.14) for displacement of each section by using the boundary conditions given below provides the dynamic solution: $w_I = 0$ and $\partial w_1/\partial x = 0$ at $x = 0, w_1 = w_2$, $\partial w_1/\partial x = \partial w_2/\partial x$, $(M_1)_1 = (M_1)_2$ and $(V_1)_1 = (V_1)_2$ on the boundary of 1 and 2-areas; $(M_1)_2 = 0$ and $(V_1)_2 = 0$ at $x = L$ (L is the total length of section and bridge); $(M_2)_i = 0$ and $(V_2)_i = 0$ at $y = \pm b_i/2$ (b_i is the width of i-area), where $(M_j)_i$ is the moment of rotation and $(V_j)_i$ is the transverse force with respect to j-axis.

The computed displacements were used to determine the strain components and then the stress components from (2). Substituting the stress components into open circuit condition enables the calculation of the ME voltage coefficient by taking into account condition that average electric field induced across the piezoelectric

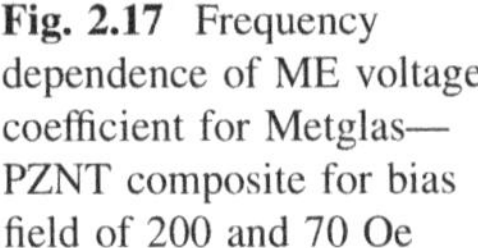

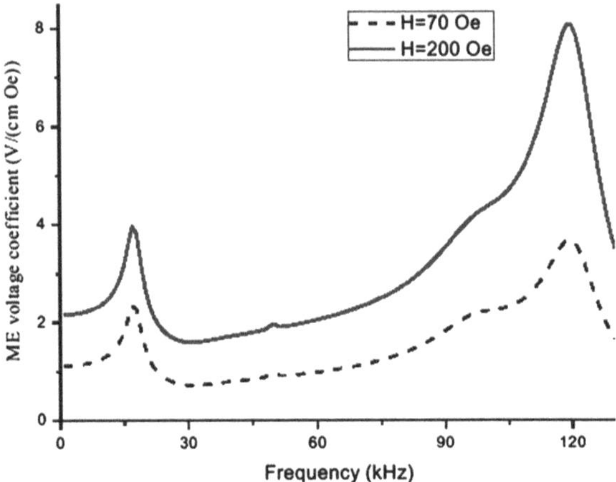

Fig. 2.17 Frequency dependence of ME voltage coefficient for Metglas—PZNT composite for bias field of 200 and 70 Oe

layer is estimated as integral of internal electric field taken over the piezoelectric thickness, $E = \frac{1}{t_p}\int_{t_p} {}^{p}E_3\,dz$. These calculations were carried out numerically. The resonance frequencies for the small rectangular section and the middle rectangular section were found to be approximately equal. The second harmonic of this mode can be seen at 97 kHz. It should be noted that theoretical estimate for the next resonance frequency (at 44 kHz) as the bending mode of large rectangular section was made using similar procedure. The bridge was not included into vibrating system since its displacement (in z direction) was negligible. Finally, the mode at 120 kHz was calculated under assumption that it came from the axial mode of the large rectangular section. The estimate for this case was found using the computing method known heretofore.

Figure 2.16 shows that the first mode at 17 kHz can be associated with simultaneous bending vibrations of the small rectangular section and the middle rectangular section, while the second mode at 44 kHz can be related to the bending of the large rectangular section. The third peak corresponds to second harmonic of first mode and occurs at 97 kHz. The fourth mode at 120 kHz came from the axial mode of the large rectangular section. Figure 2.16 reveals a frequency band of 90–130 kHz which is dependent on dc magnetic field. This band arises from the combination of the resonances. One can see two states in the ME frequency response at 70 Oe. The ME response in the frequency ranges of 25–90 kHz (state I) and 90–130 kHz (state II) is shown to be a function of frequency and the applied DC magnetic field. State I corresponds to a band lower than 1.8 V/cm Oe while State II shows a band higher than 1.8 V/cm at bias field of 70 Oe. These states are adjustable with changing H_{DC}. Some disagreement between estimates and data can be accounted for by distinctions between the geometry of actual composite structure and that of simple model used for obtaining the estimates.

Thus, we demonstrate a ME resonator exhibiting wideband behavior by fabricating a dimensionally gradient structure and combining with laminate configuration. We were able to obtain a flat ME response in the DC magnetic bias range of 52–242 Oe where the ME coefficient was higher than 3,000 mV/cm Oe under resonant condition. The promising wideband behavior as a function of frequency occurred in the range of 41–110 kHz where the ME coefficient was higher than 260 mV/cm Oe independent of applied H_{DC}. Under low H_{DC} condition, two different ME states were clearly obtained (Fig. 2.17).

2.7 Conclusions

In this section, a generalized theoretical model for low-frequency ME effects in layered composites was discussed. To describe the composite's physical properties, the exact solution of elastostatic and electrostatic equations were obtained. Expressions for the ME susceptibility and ME voltage coefficient were derived as functions of an interface coupling parameter, constituent phase material parameters, and relative volume fractions of phases. Longitudinal, transverse and in-plane cases were all considered properties. For a bilayer that is an asymmetric structure, the influence of flexural deformations of sample on ME output was estimated.

Predictions of the ME effect for various model composite systems were given including CFO–PZT and lanthanum strontium manganite–PZT. It was shown that ME effect in ferrite–PZT systems is maximum for in-plane magnetic and electric fields. The theoretical estimates of ME parameters were compared with experimental data.

The generalized theory allows for modeling of the low-frequency ME effect in bulk composites. To describe these low-frequency composite properties, an effective medium method was used. Calculation of the ME susceptibility, and ME voltage coefficient were performed as functions of volume fractions and component parameters. Composites with connectivity types 3–0 and 0–3 were considered. Larger ME coefficients were found for 3–0 composites with magnetic and/or electric fields applied along the longitudinal direction. For composites of CFO–PZT, values as high as 4 V/cm Oe were predicted for the longitudinal ME voltage coefficient. For the transverse fields orientation, ME effect was found to be in 2–3.5 times smaller than that for longitudinal orientation. Furthermore, clamping was shown to significantly reduce the ME effect.

We presented a theory for the resonance enhancement of ME interactions at frequencies corresponding to EMR. Frequency dependence for ME voltage coefficients are obtained using the simultaneous solution of electrostatic, magnetostatic, and elastodynamic equations. The ME effect at bending mode in a bilayer is shown to be dependent on boundary conditions. A giant ME interaction at the lowest frequency is predicted for a bilayer fixed at one end and is free to vibrate at the other end. The ME voltage coefficients are estimated from known material parameters (piezoelectric coupling, magnetostriction, elastic constants, etc.) of composite components. It is shown that the ME coupling in the EMR region exceeds the low-frequency value by more than an order of magnitude.

References

Bichurin MI, Petrov VM, Srinivasan G (2002a) Modelling of magnetoelectric effect in ferromagnetic/piezoelectric multilayer composites. Ferroelectrics 280:165

Bichurin MI, Petrov VM, Srinivasan G (2002b) Theory of low-frequency magnetoelectric effects in ferromagnetic-ferroelectric layered composites. J Appl Phys 92:7681

Bichurin MI, Petrov VM, Srinivasan G (2003) Theory of low-frequency magnetoelectric coupling in magnetostrictive-piezoelectric bilayers. Phys Rev B 68:054402

Bichurin MI, Petrov VM, Srinivasan G (2009) Low-frequency magnetoelectric effects in ferrite–piezoelectric nanostructures. J Magn Magn Mater 321:846–849

Gheevarughese V, Laletsin U, Petrov VM, Srinivasan G, Fedotov NA (2007) Low-frequency and resonance magnetoelectric effects in lead zirconate titanate and single-crystal nickel zinc ferrite bilayers. J Mater Res 22:2130–2135

Harshe G, Dougherty JO, Newnham RE (1993a) Theoretical modelling of multilayer magnetoelectric composites. Int J Appl Electromagn Mater 4:145

Harshe G, Dougherty JP, Newnham RE (1993b) Theoretical modelling of 3–0, 0–3 magnetoelectric composites. Int J Appl Electromagn Mater 4:161

Mandal SK, Sreenivasulu G, Petrov VM, Srinivasan G (2011) Magnetization-graded multiferroic composite and magnetoelectric effects at zero bias. Phys Rev B 84:014432

Newnham RE, Skinner DP, Cross LE (1978) Connectivity and piezoelectric-pyroelectric composites. Mater Res Bull 13:525

Osaretin IA, Rojas RG (2010) Theoretical model for the magnetoelectric effect in magnetostrictive/piezoelectric composites. Phys Rev B 82:174415

Park C-S, Avirovik D, Bichurin MI, Petrov VM, Priya S (2012) Tunable magnetoelectric response of dimensionally gradient laminate composites. Appl Phys Lett 100:212901

Petrov VM, Srinivasan G (2008) Enhancement of magnetoelectric coupling in functionally graded ferroelectric and ferromagnetic bilayers. Phys Rev B 78:184421

Petrov M, Bichurin MI, Laletin VM, Paddubnaya N, Srinivasan G (2004) Modeling of magnetoelectric effects in ferromagnetic/piezoelectric bulk composites. In: Fiebig M, Eremenko VV, Chupis IE (eds) Magnetoelectric interaction phenomena in crystals-NATO science series II, vol 164. Kluwer Academic Publishers, London, pp 65–70

Petrov VM, Srinivasan G, Laletsin U, Bichurin MI, Tuskov DS (2007) Magnetoelectric effects in porous ferromagnetic-piezoelectric bulk composites: experiment and theory. Phys Rev B 75:174422

Petrov VM, Srinivasan G, Bichurin MI, Galkina TA (2009) Theory of magnetoelectric effect for bending modes in magnetostrictive-piezoelectric bilayers. J Appl Phys 105:063911

Chapter 3
Maxwell-Wagner Relaxation in ME Composites

Abstract The main objective is relaxation processes in ME composites. For layered ferrite–piezoelectric composite, the Maxwell-Wagner relaxation of ME susceptibility and ME voltage coefficient is found to have Debay character and to be normal for ME susceptibility and inverse for ME voltage coefficient. In ferrite–piezoelectric bulk composite, the Maxwell-Wagner relaxation of ME susceptibility and ME voltage coefficient is inverse for ME voltage coefficient and can be both normal and inverse for ME susceptibility.

Ferrite–piezoelectric composites can be considered as a thermodynamic system that are capable of responding to an infinitesimal externally applied field. Relaxation can then be considered as the self-adjustment of the thermodynamic system in response to an external force (Novick and Berry 1972) that maintains equilibrium. It reflects an adjustment of internal parameters to the balanced values. Specifically, dielectric relaxation is a time-dependent induced polarization P response to an applied electric field E. The process is characterized by dispersion in the internal parameter $\delta P/\delta E$. Polar dielectrics with low resistivities often have a strongly enhanced permittivity at low frequencies, due to charge separation. Such space charge polarization contributions to the permittivity are characterized by low-frequency dispersion (Petrov et al. 2004, 2007). Investigations have shown that the effects of space charge polarization relaxation reveals itself not only in the dielectric permittivity, but also in the elastic constants of layered polar dielectrics.

In the case of dielectric relaxation in the quasistatic frequency range, when the applied electric field duration period may exceed times of seconds or more, space charge polarization contributions to the dielectric permittivity have sufficient time to respond to the drive: consequently, the permittivity has a constant value. However, as the frequency is increased, this polarization mechanism cannot respond efficiently to the drive, rather there is a phase angle between drive and response. The tangent of this phase angle is the dielectric loss factor. Thus, the material is characterized by the complex dielectric permittivity that depends on frequency

M. Bichurin and V. Petrov, *Modeling of Magnetoelectric Effects in Composites*, Springer Series in Materials Science 201, DOI: 10.1007/978-94-017-9156-4_3,

$$\varepsilon_{33} = \varepsilon - i\gamma/\omega \tag{3.1}$$

where $\varepsilon'(\omega)$ is the real part (i.e., the permittivity of storage compliance), and $\varepsilon''(\omega)$ is imaginary part (i.e., the phase angle or energy loss). Dynamic experiments in which the electric field applied to the material is periodically varied with a circular frequency are standard methods for measuring the dielectric constant. There are many impedance analyzers capable of measuring the complex frequency-dependent dielectric constant over frequencies extending from micro to mega Hertz.

Since dynamic response functions allow for the description of frequency-dependent materials parameters, they are suitable for use in the analysis of the ME susceptibility and voltage coefficient of ferrite–piezoelectric composites, both of which are complex functions of frequency. Investigations of the relaxation phenomenon of ferrite–piezoelectric composites have previously been reported.

The relations that we obtained for the effective parameters of composite in the last section did not consider their frequency dependence. Calculations were done using the ferroelectric polarization, which will provide a fair estimate of the high-frequency polarization. Accumulation of electric charges at the boundaries of composite components will result in additional contributions to the polarization: albeit low-frequency ones. Said accumulation of free charges at the interfaces between component phases will result in dielectric dispersion and losses under low-frequency alternating fields, which is known as Maxwell-Wagner relaxation. The purpose of the following section is an analysis of Maxwell-Wagner relaxation of the effective parameters in ferrite–piezoelectric composites, in particular, the ME susceptibility and voltage coefficient.

3.1 Layered Composites

Now, let us consider Maxwell-Wagner relaxation of the ME parameters for a multilayer composite with a 2–2 type connectivity (Petrov et al. 2004; Bichurin et al. 2012). Following from 3.1, the specific equations, in view of finite electrical conductivities of component phases, for the complex permittivity of the components is

$$\begin{aligned} {}^{p}\varepsilon_{33} &= {}^{p}\varepsilon - i^{p}\gamma/\omega, \\ {}^{m}\varepsilon_{33} &= {}^{m}\varepsilon - i^{m}\gamma/\omega; \end{aligned} \tag{3.2}$$

where ${}^{p}\gamma$ and ${}^{m}\gamma$ are the conductivities of the piezoelectric and magnetostrictive phases; and ω the circular frequency, where the frequency is less than that of electromechanical resonance. The effective parameters of the composite can be then determined by an average of expressions for the components of strain, electric, and magnetic inductions. Furthermore, assume that the component phase layers are thin and arranged in the plane OX_1X_2; that the piezoelectric component is polarized along the OX_3 axis, along which same axis the an electric field with a circular frequency ω is applied; and that a magnetic bias and variable magnetic fields are applied along the OX_1 axis. I

The general characteristics of the frequency-dependent ME susceptibility will fulfill the Debye formula:

$$\begin{aligned} &\alpha_{13} = \alpha'_{13} - i\alpha''_{13}, \\ &\alpha'_{13} = \alpha_{13\infty} + \Delta\alpha_{13}/(1+\omega^2\tau_\alpha^2);\ \alpha''_{13} = \Delta\alpha_{13}\omega\tau_\alpha/(1+\omega^2\tau_\alpha^2); \end{aligned} \tag{3.3}$$

where $\Delta\alpha_{13} = \alpha_{130} - \alpha_{13\infty}$ is the relaxation strength, α_{130} and $\alpha_{13\infty}$ are the static ($\omega \to 0$) and high-frequency ($\omega \to \infty$) ME susceptibilities, and τ_α the relaxation time. The transverse static and high-frequency ME susceptibilities, and in addition the relaxation time, can be found from the solutions of 2.5–2.8. Then, by supposing that the symmetry of the piezoelectric phase is ∞m, and that the magnetic phase possesses cubic symmetry, we obtain the following expressions:

$$\begin{aligned} \alpha_{130} &= \frac{{}^m\gamma\upsilon k(1-\upsilon)({}^mq_{12}+{}^mq_{11}){}^pd_{31}}{[{}^p\gamma(1-\upsilon)+{}^m\gamma\upsilon][({}^ms_{12}+{}^ms_{11})\upsilon+({}^ps_{11}+{}^ps_{12})(1-\upsilon)]}, \\ \alpha_{13\infty} &= \frac{{}^m\varepsilon\upsilon k(1-\upsilon)({}^mq_{12}+{}^mq_{11})}{\left[[\upsilon{}^m\varepsilon+{}^p\varepsilon(1-\upsilon][({}^ms_{12}+{}^ms_{11})\upsilon+({}^ps_{11}+{}^ps_{12})(1-\upsilon)]-2{}^pd_{31}^2(1-\upsilon)^2\right]}, \end{aligned} \tag{3.4}$$

As an example, we shall consider the composite to consist of polarized ferroelectric PZT ceramics and nickel ferro-spinel. In numerical calculations, the following values of the materials parameters for composite phase components can be used:
${}^ps_{11} = 15.3 \times 10^{-12}$ m²/N, ${}^ps_{12} = -5 \times 10^{-12}$ m²/N, ${}^ps_{13} = -7.22 \times 10^{-12}$ m²/N, ${}^ps_{33} = 17.3 \times 10^{-12}$ m²/N, ${}^ms_{11} = 15.3 \times 10^{-12}$ m²/N, ${}^ms_{12} = -5 \times 10^{-12}$ m²/N, ${}^mq_{33} = -1880 \times 10^{-12}$ m/A, ${}^mq_{31} = 556 \times 10^{-12}$ m/A, ${}^pd_{31} = -175 \times 10^{-12}$ m/V, ${}^pd_{33} = -400 \times 10^{-12}$ m/V, ${}^m\mu_{33}/\mu_0 = 3$, ${}^p\varepsilon/\varepsilon_0 = 1750$, ${}^m\varepsilon/\varepsilon_0 = 10$, ${}^m\gamma = 10^{-5}$ (Ohm m)$^{-1}$, ${}^p\gamma = 10^{-13}$ (Ohm m)$^{-1}$.

Relaxation of the effective permittivity of the composite was then calculated, and is shown in Fig. 3.1. Under the conditions of ${}^p\gamma/{}^m\gamma << 1$, ${}^p\varepsilon/{}^m\varepsilon >> 1$, and $\upsilon << 1$, it is possible to obtain a giant increase in the static dielectric permittivity. The results are in agreement with prior reports for ferroelectric–polymer composites. In the work given in Fig. 3.1, the dependence of the parameters for dielectric relaxation depended on an interface connection parameter. The strength of the relaxation was found to decrease with increase of this interphase–interface parameter.

Figure 3.2 illustrates the dependence of the relaxation of the ME susceptibility on piezoelectric phase volume fraction depth. A large relaxation strength is

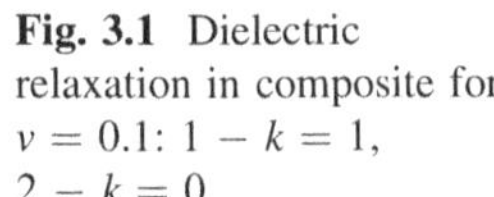

Fig. 3.1 Dielectric relaxation in composite for $v = 0.1$: $1 - k = 1$, $2 - k = 0$

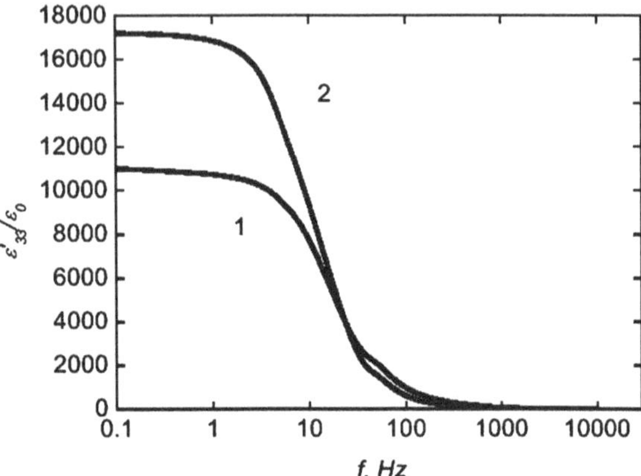

Fig. 3.2 Frequency dependence real (1, 3) and imaginary (2, 4) parts of a ME susceptibility: 1, $2 - v = 0.001$, 3, $4 - v = 0.9$

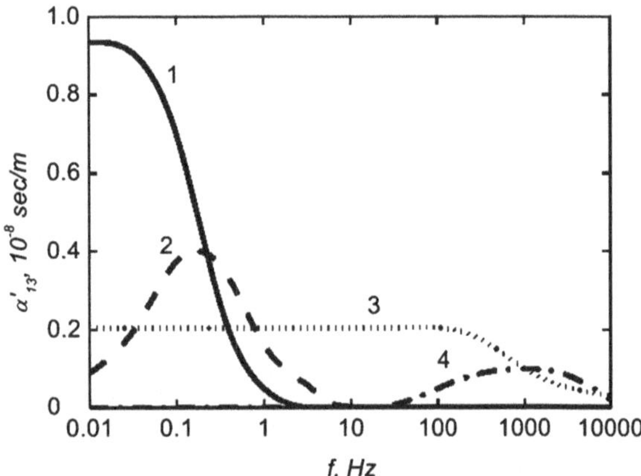

characteristic for a composite whose piezoelectric phase component has a big permittivity and whose ferrite one is electrically conductive. In the case of ${}^m\varepsilon/{}^p\varepsilon << 1$ and ${}^p\gamma/{}^m\gamma << 1$, a maximum value of the relaxation strength is observed at a piezoelectric phase volume fraction of $v \approx [({}^p s_{11} +{}^p s_{12})/({}^m s_{11} +{}^m s_{12})]^{1/2} \, ({}^p\gamma/{}^m\gamma)^{1/2}$. If the compliances of both composite phase components are taken as equal, then one gets $\upsilon \approx ({}^p\gamma/{}^m\gamma)^{1/2}$.

Relaxation of the ME susceptibility is shown in Fig. 3.2. For ${}^p\gamma/{}^m\gamma << \upsilon << 1$, the static ME susceptibility approaches a value equal to $({}^m q_{12} + {}^m q_{11})\,{}^p d_{31}/({}^p s_{11} + {}^p s_{12})$. For a composite with a composition dictated by the condition $\upsilon_1 = ({}^p h/{}^m h)^{1/2} \approx 10^{-4}$ where ${}^p h$ and ${}^m h$ are the thicknesses of the piezoelectric and ferrite phase components, the value of the ME susceptibility is equal $0.94 \cdot 10^{-8}$s/m s/m. This large magnitude of the static ME susceptibility is due to a large local electric field in the thin piezoelectric phase component, to the significant electrical conductance of the ferrite layer, and to the large internal mechanical

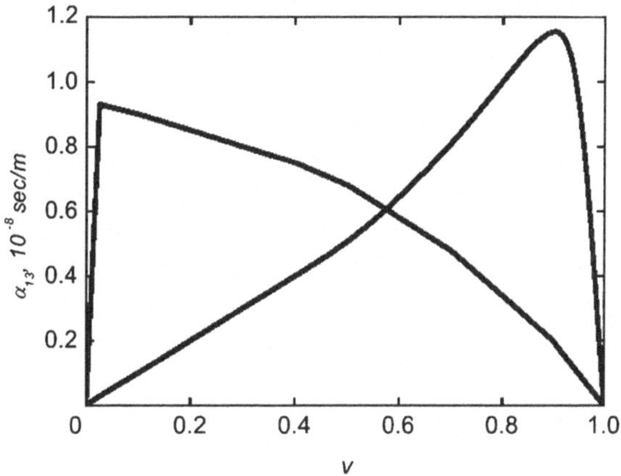

Fig. 3.3 Concentration dependence of static and high-frequency magnetoelectric susceptibilities

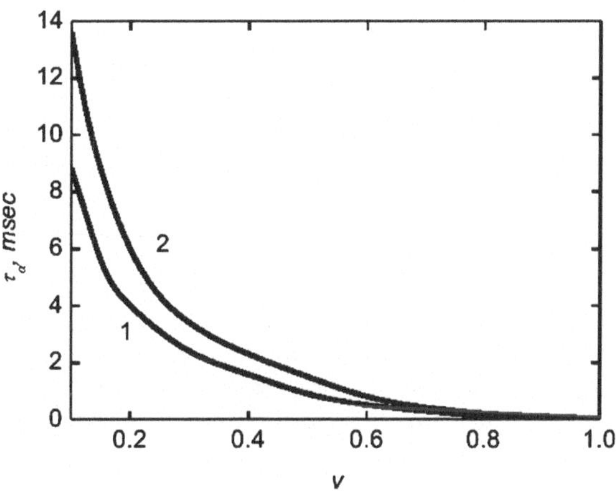

Fig. 3.4 Concentration dependence of relaxation time of ME susceptibility: $1 - k = 1$, $2 - k = 0.1$

stresses ${}^pT_j$ and ${}^mT_j$ ($j = 1{,}2$) that are induced by the electrical field in the piezoelectric component. The predicted maximum value for the ME susceptibility of this composite exceeds the experimentally observed one for known materials parameters (Fig. 3.3).

In the case of a weak piezoelectric effect considered here as $({}^pd_{31}^2/[({}^ps_{11} + {}^ps_{12})\,{}^p\varepsilon] << 1)$, the relaxation time as follows from 3.4 at ${}^p\gamma/{}^m\gamma << \upsilon(1-\upsilon)$ is controlled mainly by the charging time of the piezoelectric layer capacitance, across the resistance of the ferrite layer: $\tau_\alpha \approx ({}^p\varepsilon/{}^m\gamma)(1-\upsilon)/\upsilon$, as shown in Fig. 3.4. With increase of the piezoelectric phase volume fraction, the relaxation time decreases rapidly, whereas with increase of the piezoelectric constant of the ferroelectric phase, that of the composite also increases. The stresses ${}^pT_1$ and ${}^pT_2$ (caused by a transverse piezoelectric effect) induce additional charge on the piezoelectric layer, increase the charging time of its capacity, and decrease the relaxation time of the ME susceptibility.

Next, we consider the effect that the interfacial parameter k has on the relaxation time. As follows from 3.4, the relaxation time τ_α and relaxation frequency $\omega_r = 1/\tau_\alpha$ (determined by the frequency at which maximum in the imaginary ME susceptibility occurs) can be predicted for a wide range of volume fractions of the component phases, and also for various choices of materials for the component phases. Since the frequency dependence of the transverse ME voltage coefficient is defined by the Debye formulas, we can arrive at relations analogously to 3.4 which provides us the dependence of the relaxation time on various parameters, as follows:

$$\begin{aligned}\alpha_{E,T} &= \alpha'_{E,T} - i\alpha''_{E,T}, \\ \alpha'_{E,T} &= \alpha_{E,T\,\infty} + \Delta\alpha_{E,T}/(1+\omega^2\tau^2_{\alpha T}); \\ \alpha''_{E,T} &= \Delta\alpha_{E,T}\omega\tau_{\alpha T}/(1+\omega^2\tau^2_{\alpha T});\end{aligned} \tag{3.5}$$

where

$$\Delta\alpha_{E,T} = -\alpha_{E,T\,\infty} = \frac{v^p d_{31} k(1-v)(^m q_{12} + ^m q_{11})}{[^m s' v + ^p s'(1-v)] - 2^p d^2_{31}\,(1-v)},$$

$$\tau_{\alpha T} = \frac{^p\varepsilon}{^p\gamma} - \frac{2k(1-v)^p d^2_{31}}{^p\gamma[v^m s' + ^p s'(1-v)]}$$

with $^m s' = {}^m s_{11} + {}^m s_{12}$ and $^p s' = {}^p s_{11} + {}^p s_{12}$.

Following this relationship, we can see in Fig. 3.4 that increasing k has the same effect on relaxation time as previously shown for increasing piezoelectric constant: it increases notably the value of τ_α.

Unlike the ME susceptibility, the real part of the ME voltage coefficient increases with increasing frequency: i.e., inverse relaxation occurs, as shown in Fig. 3.5. The relaxation strength was maximum for $\upsilon \approx 0.5$. In the case of a weak piezoelectric effect (defined as $d^2_{31}/[(^p s_{11} + {}^p s_{12})\,{}^p\varepsilon] << 1$), the relaxation time is defined controlled by the discharge time of the piezoelectric layer capacitance through its own resistance $\tau_{\alpha T} \approx {}^p\varepsilon/{}^p\gamma$, and does not depend on volume fraction of the component phases. In this case, it can be shown that $\tau_{\alpha T} >> \tau_\alpha$, as illustrated in Fig. 3.6. This figure also shows that increasing piezoelectric phase volume fraction decreases $\tau_{\alpha T}$. This diminution is due to the appearance of additional charge on the piezoelectric layer capacitance under effect of stresses $^pT_1$ and $^pT_2$, which is induced by external magnetic fields applied to the composite.

Thus, in layered ferrite–piezoelectric composites, a strong relaxation of the ME susceptibility is observed, whereas, for the ME voltage coefficient, a strong inverse effect occurs. The relaxation time and relaxation frequency of the ME susceptibility can be changed over a wide range of values, by varying the volume fractions of the composite's component phases, and also by selection of the initial component phases as different piezoelectric and magnetostrictive phases can have different values of property tensors.

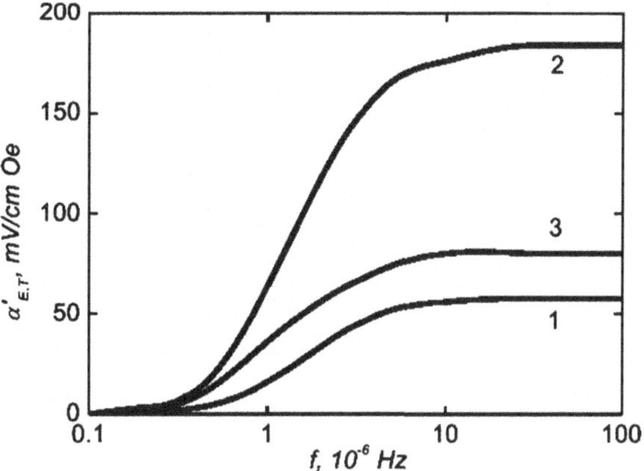

Fig. 3.5 Frequency dependence of real part of ME voltage coefficient: $1 - \nu = 0.1$, $2 - \nu = 0.5$, $3 - \nu = 0.9$

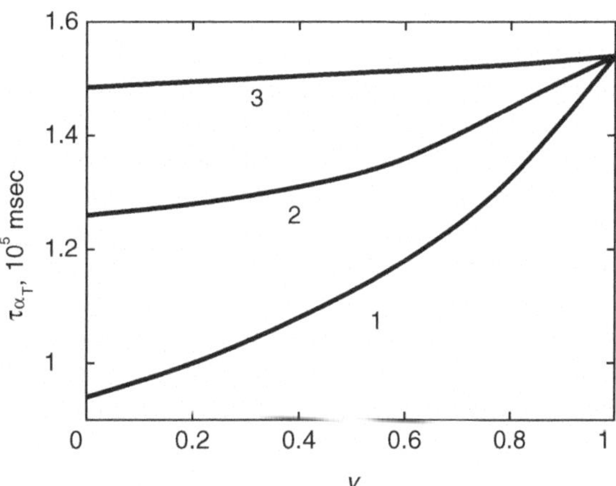

Fig. 3.6 Concentration dependence of relaxation time of ME voltage coefficient: $1 - k = 1$, $2 - k = 0.5$, $3 - k = 0.1$

3.2 Bulk Composites

As an example, consider the cubic model of bulk ferrite–piezoelectric composites (Petrov et al. 2004) in which one phase has connectivity in all three directions and in which a second isolated phase does not have connectivity in any direction. Following known classifications (Newnham et al. 1978), this specified composite has a connectivity of the type 3–0. For the piezoelectric and magnetostrictive phases, the equations for the strain, dielectric displacement, and the magnetic induction were presented earlier. In addition, in view of a finite electrical conductivity, the permittivities of the component phases are given by 3.2.

The general formulas for defining the effective parameters of the composite can then be obtained by averaging the expressions for components of the strains, and the electric and magnetic inductions following 2.5–2.8. Then assume that the sample has the form of a disk that is oriented in the plane OX_1X_2, where the

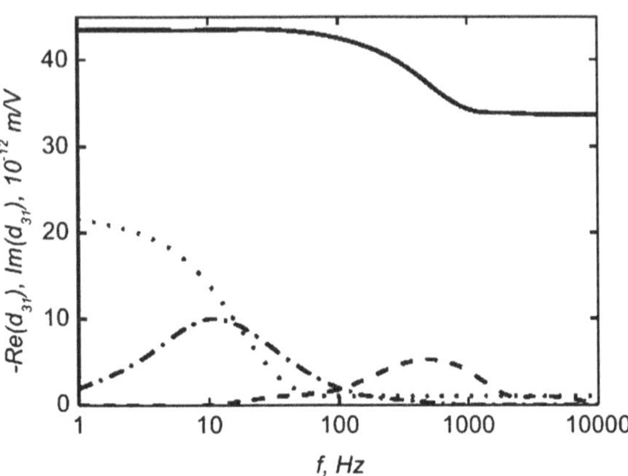

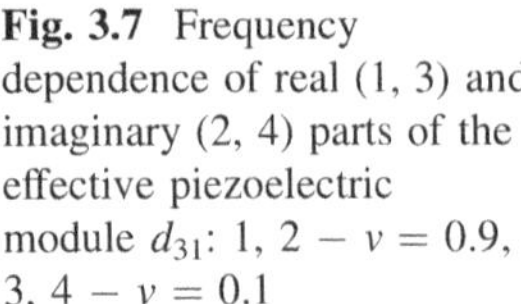
Fig. 3.7 Frequency dependence of real (1, 3) and imaginary (2, 4) parts of the effective piezoelectric module d_{31}: 1, 2 – $v = 0.9$, 3, 4 – $v = 0.1$

piezoelectric phase is polarized along the axis OX_3, an electric field with a frequency ω is applied along the same axis, and both a magnetic bias and an alternate magnetic field are applied along axis OX_1.

Static and high-frequency ME susceptibilities, in addition to the relaxation time, can be obtained from solutions to 3.4 by applying boundary conditions for strains, stresses, and electric and magnetic fields. Then, suppose that the symmetry of the piezoelectric phase is ∞m, and that the magnetic phase possesses cubic symmetry. An analytical expression can then be obtained, but due to inconvenience of its form, solution of equations can only be fulfilled numerically (Petrov et al. 2007). As an example system to apply this numerical solution, we shall consider a composite consisting of poled PZT ferroelectric ceramics and nickel ferro-spinel, which has the following values of component phase material parameters: ${}^{p}s_{11} = 15.3 \times 10^{-12}$ m^2/N, ${}^{p}s_{12} = -5 \times 10^{-12}$ m^2/N, ${}^{p}s_{13} = -7.22 \times 10^{12}$ m^2/N, ${}^{p}s_{33} = 17.3 \times 10^{12}$ m^2/N, ${}^{m}s_{11} = 15.3 \times 10^{12}$ m^2/N, ${}^{m}s_{12} = -5 \times 10^{-12}$ m^2/N, ${}^{m}q_{33} = -1880 \times 10^{-12}$ m/A, ${}^{m}q_{31} = 556 \times 10^{-12}$ m/A, ${}^{p}d_{31} = -50 \times 10^{-12}$ m/V, ${}^{p}d_{33} = -100 \times 10^{-12}$ m/V, ${}^{m}\mu_{33}/\mu_0 = 2$, ${}^{p}\varepsilon/\varepsilon_0 = 1{,}000$, ${}^{m}\varepsilon/\varepsilon_0 = 10$, ${}^{m}\gamma = 10^{-5}$ (Ohm·m)$^{-1}$, ${}^{p}\gamma = 10^{13}$ (Ohm·m)$^{-1}$.

In Figs. 3.7 and 3.8, the frequency dependences of the effective piezoelectric modules of this 0–3 composite is illustrated. With increase of the piezoelectric phase volume fraction, the relaxation strength increases, and the relaxation frequency decreases. Thus, parameters of the piezoelectric relaxation can be varied by changing the relative volume fractions of the composite component phases.

The relaxation of the effective dielectric permittivity is illustrated in Fig. 3.9. Under the conditions of ${}^{p}\gamma/{}^{m}\gamma << 1, {}^{p}\varepsilon/{}^{m}\varepsilon >> 1$, and $\upsilon << 1$, pronounced increases in the static permittivity can be seen, similar to that found earlier for layered composites. In this case, the relaxation frequency does not depend much on the volume fraction of the components, and is mainly defined by electric parameters. Interestingly, as shown in Fig. 3.10, there is a frequency for which the effective permittivity does not depend much on the piezoelectric phase volume fraction.

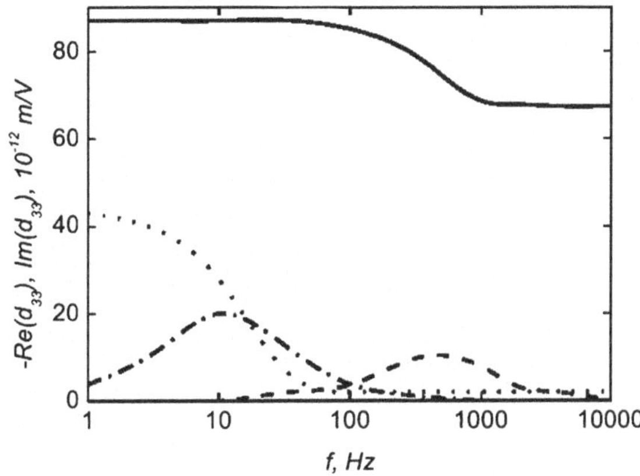

Fig. 3.8 Frequency dependence of real (1, 3) and imaginary (2, 4) parts of the effective piezoelectric module d_{33}: 1, 2 – $v = 0.9$, 3, 4 – $v = 0.1$

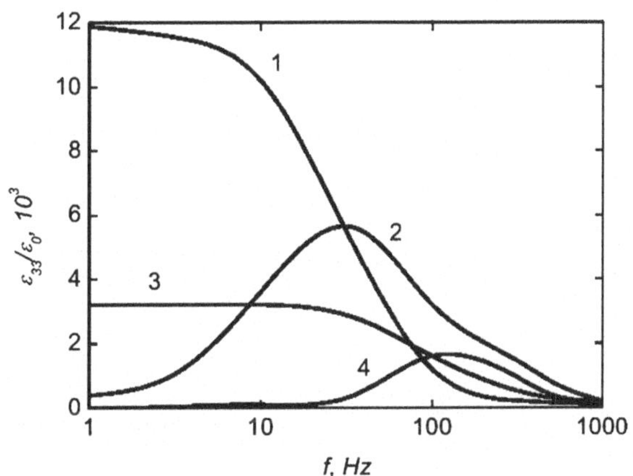

Fig. 3.9 Frequency dependence of real (1, 3) and imaginary (2, 4) parts of effective permittivity: 1, 2 – $v = 0.1$, 3, 4 – $v = 0.9$

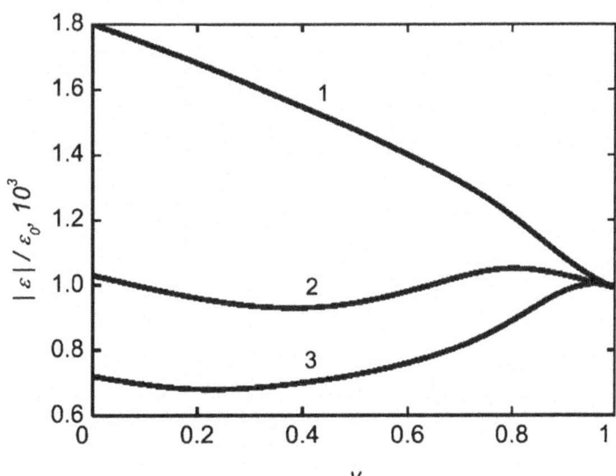

Fig. 3.10 Concentration dependence of effective permittivity module: 1 – $f = 200$ Hz, 2 – $f = 350$ Hz, 3 – $f = 500$ Hz

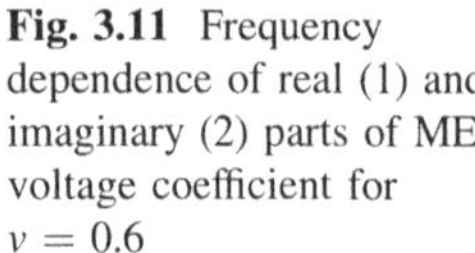
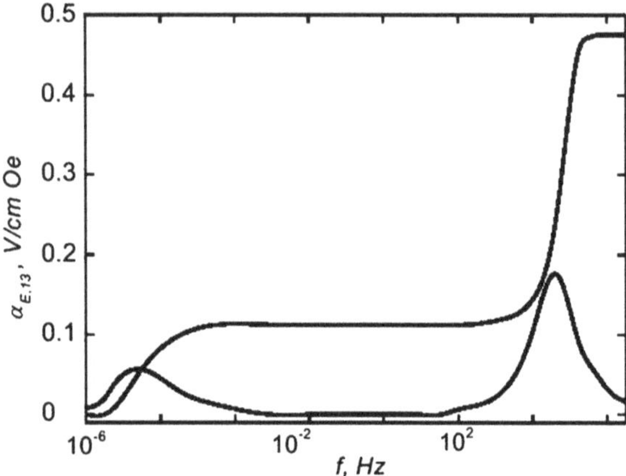

Fig. 3.11 Frequency dependence of real (1) and imaginary (2) parts of ME voltage coefficient for $v = 0.6$

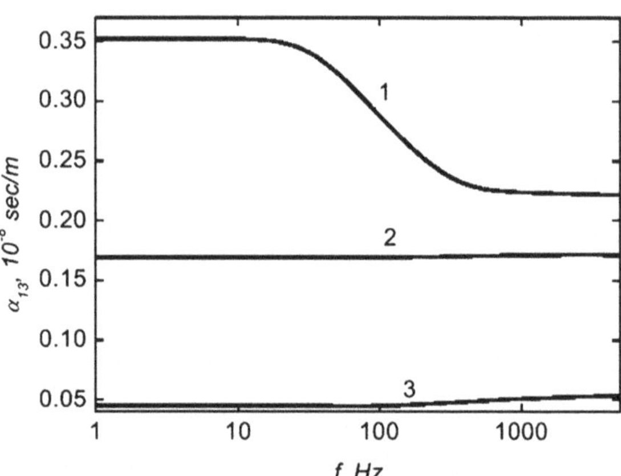

Fig. 3.12 Frequency dependence of ME susceptibility real part: $1 - v = 0.5$, $2 - v = 0.8$, $3 - v = 0.95$

The frequency dependence of the ME voltage coefficient is shown in Fig. 3.11. The real part of the ME voltage coefficient can be seen to increase with increasing frequency, i.e., an inverse relaxation occurs. The relaxation spectrum is characterized by two regions of dispersion. The first cover the range of infra-low frequency. The relaxation time for this region depends on weakly on the component phase volume fractions, and is controlled by the discharge time of the piezoelectric phase capacitance through its own resistance. The second region of relaxation covers the frequency range over which the voltages on capacitance elements 2, 3, and 4 are equalized. In this region, the relaxation is defined by the charging time of the capacitance of element 2 through the resistance of the ferrite phase component, and it varies as the relative volume fraction of the phases is changed.

Relaxation of the ME susceptibility is then shown in Fig. 3.12. The relaxation time of the ME susceptibility is defined by the charging time of the capacitance of

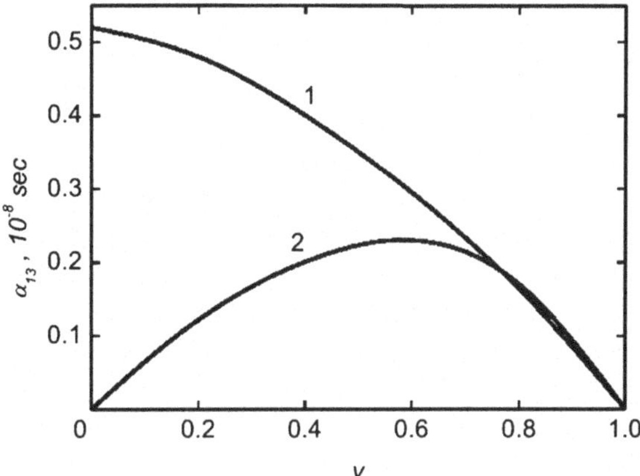

Fig. 3.13 Dependence of static (1) and high-frequency (2) ME susceptibilities on volume fraction of the piezoelectric component

element 2, through the resistance of the ferrite component. The relaxation strength decreases with increase of the piezoelectric component volume fraction, and for volumes of $\upsilon > 0.8$ becomes negative. Thus, relaxation of the ME susceptibility depends on the volume fraction of component phases, and can be either direct or inverse. Interestingly, there is a specific value for the piezoelectric phase volume fraction for which there is no frequency dependence of the ME susceptibility. This specific value is defined by composite component parameters, as illustrated in Fig. 3.12. The static and high-frequency ME susceptibilities are then shown in Fig. 3.13. It can be seen that the static ME susceptibility has a peak for $\upsilon << 1$.

3.3 Conclusions

In this chapter, a generalized theoretical model for low-frequency ME effects in layered composites was introduced. To describe the composite's physical properties, effective parameters were derived. This model allowed us to define on the basis of an exact solution the effective mechanical, electric, magnetic, and ME parameters of layered ferrite–piezoelectric composite. Expressions for the effective parameters, (including ME susceptibility and ME voltage coefficient) were derived as functions of an interface coupling parameter, constituent phase material parameters, and relative volume fractions of phases. Longitudinal, transverse and in-plane cases were all considered.

The approach predicted giant ME effects in ferrite–PZT ceramic composites. Predictions of the ME effect for various model ceramic composite systems were given including CFO–BTO, CFO–PZT, NFO–PZT, and lanthanum strontium manganite–PZT. It was shown that ME effect in ferrite–PZT systems is maximum for magnetic and electric fields applied along an in-plane orientation: however, this has yet to be experimentally verified. Using the model, a comparison of the ME parameters was then made between calculated values and experimental data.

From this comparison, the importance of an interfacial coupling parameter between phases was inferred. This interphase interfacial connection parameter was shown to be weak for CFO–PZT and lanthanum strontium manganite–PZT composites, but near ideal for NFO–PZT.

In addition, the generalized theory allows for modeling of the low-frequency ME effect in bulk composites. To describe these low-frequency composite properties, an effective medium method was used. Calculation of the effective permittivity and permeability, piezoelectric and piezomagnetic modules, ME susceptibility, and ME voltage coefficient were performed as functions of volume fractions and component parameters. Composites with dimensional connectivities of types 3–0 and 0–3 were considered.

Larger ME coefficients were found for 3–0 composites with magnetic and/or electric fields applied along the longitudinal direction. For composites of CFO–PZT, values as high as 4 V/(cmOe) were predicted for the ME voltage coefficient. Fields applied along transverse orientations were found to have ME effects about 2–3.5× smaller than those for longitudinally applied fields. Furthermore, clamping was shown to significantly reduce the ME effect. Finally, in bulk ferrite–piezoelectric composites, large relaxation effects in the ME susceptibility were found with increasing frequency. The relaxation time and frequency of the ME susceptibility and voltage coefficient can be varied by changing the volume fraction of component phases or by changes in the composite component parameters.

References

Bichurin M, Petrov V, Priya S, Bhalla A (2012) Multiferroic magnetoelectric composites and their applications. Adv Condens Matter Phys Article ID 129794

Newnham RE, Skinner DP, Cross LE (1978) Connectivity and piezoelectric-pyroelectric composites. Mater Res Bull 13:525

Novick A, Berry B (1972) Anelastic relaxation in crystalline solids. Academic Press, New York, p 677

Petrov VM, Bichurin MI, Srinivasan G (2004a) Maxwell-Wagner relaxation in magnetoelectric composites. Tech Phys Lett 30:81

Petrov VM, Bichurin MI, Srinivasan G, Zhai J, Viehland D (2007) Dispersion characteristics for low-frequency magnetoelectric coefficients in bulk ferrite-piezoelectric composites. Solid State Commun 142:515

Petrov VM, Bichurin MI, Laletin VM, Paddubnaya N, Srinivasan G (2004b) Modeling of magnetoelectric effects in ferromagnetic/piezoelectric bulk composites. In: Fiebig M, Eremenko VV, Chupis IE (eds) Magnetoelectric interaction phenomena in crystals-NATO science series II, vol 164. Kluwer Academic Publishers, London, pp 65–70

Chapter 4
Magnetoelectric Effect in Electromechanical Resonance Region

Abstract We present a theory for the resonance enhancement of ME interactions at frequencies corresponding to electromechanical resonance (EMR). Frequency dependence of ME voltage coefficients are obtained using the simultaneous solution of electrostatic, magnetostatic and elasto-dynamic equations. The ME voltage coefficients are estimated from known material parameters (piezoelectric coupling, magnetostriction, elastic constants, etc.) of composite components. It is shown that the resonance enhancement of ME interactions is observed at frequencies corresponding to EMR and ME coupling in the EMR region exceeds the low-frequency value by more than an order of magnitude. It was found that the peak transverse ME coefficient at EMR is larger than the longitudinal one. The results of calculations obtained for a nickel-ferrite spinel–PZT composite are in good agreement with the experimental data.

The magnetoelectric (ME) effect in composites is caused by mechanically coupled magnetostrictive and piezoelectric subsystems: it is present in neither subsystem separately. Under magnetic field owing to the magnetostriction of the ferrite component, there are stresses which are elastically transmitted in the piezoelectric phase resulting in polarization changes via piezoelectricity. Because the ME effect in composites is due to mechanically coupled piezoelectric and magnetostrictive subsystems, it sharply increases in the vicinity of the electromechanical resonance (EMR) frequency (Bichurin et al. 2010, 2012a, b; Filippov et al. 2004; Zhang et al. 2009).

4.1 Modeling of Magnetoelectric Effect at Longitudinal and Radial Modes

In this section, we present a theory for the resonance enhancement of magnetoelectric (ME) interactions at frequencies corresponding to electromechanical resonance (EMR). Frequency dependence of ME voltage coefficients are obtained

M. Bichurin and V. Petrov, *Modeling of Magnetoelectric Effects in Composites*, Springer Series in Materials Science 201, DOI: 10.1007/978-94-017-9156-4_4,

using the simultaneous solution of electrostatic, magnetostatic, and elastodynamic equations. The ME voltage coefficients are estimated from known material parameters (piezoelectric coupling, magnetostriction, elastic constants, etc.) of composite components. It is shown that the resonance enhancement of ME interactions is observed at frequencies corresponding to EMR and ME coupling in the EMR region exceeds the low-frequency value by more than an order of magnitude. It was found that the peak transverse ME coefficient at EMR is larger than the longitudinal one. The results of calculations obtained for a CFO–PZT composite are in good agreement with the experimental data.

Mechanical oscillations of a ME composite can be induced either by alternating magnetic or electric fields. If the length of the electromagnetic wave exceeds the spatial size of the composite by some orders of magnitude, then it is possible to neglect gradients of the electric and magnetic fields within the sample volume. Therefore, based on elastodynamics and electrostatics, the equations of medium motion are governed by

$$\bar{\rho}\frac{\partial^2 u_i}{\partial t^2} = V\frac{\partial^p T_{ij}}{\partial x_j} + (1-V)\frac{\partial^m T_{ij}}{\partial x_j}, \tag{4.1}$$

where u_i is the displacement vector component, $\overline{\rho} = V^p\rho + (1-V)^m\rho$ is the average mass density, V is the ferroelectric volume fraction, ${}^p\rho$ and ${}^m\rho$, ${}^pT_{ij}$, and ${}^mT_{ij}$ are the density and stress tensor components of ferroelectric and ferromagnetic phase, correspondingly.

Simultaneous solution of elasticity equations and 4.1 by use of appropriate boundary conditions allows one to find the ME voltage coefficient. Since the solution depends on the composite shape and orientation of applied electric and magnetic fields, we consider some of the most general cases in this chapter.

4.1.1 Narrow Composite Plate

First, let us consider a composite that has the form of a narrow plate which has a length L, as shown in Fig. 4.1 (Filippov et al. 2004; Bichurin et al. 2003, 2010).

Tensorial expressions for the strain in the ferromagnetic layers ${}^mS_i$, and the strain ${}^pS_i$ and electrical displacement D in the ferroelectric layers have the form for the bias field directed perpendicular to the sample plane (along z-axis):

$$\begin{aligned} {}^pS_1 &= {}^ps_{11}{}^pT_1 + {}^ps_{12}{}^pT_2 + {}^pd_{31}E_3, \\ {}^pS_2 &= {}^ps_{12}{}^pT_1 + {}^ps_{11}{}^pT_2 + {}^pd_{31}E_3, \end{aligned} \tag{4.2}$$

$$\begin{aligned} {}^mS_1 &= {}^ms_{11}{}^mT_1 + {}^ms_{12}{}^mT_2 + {}^mq_{31}H_3, \\ {}^mS_2 &= {}^ms_{12}{}^mT_1 + {}^ms_{11}{}^mT_2 + {}^mq_{31}H_3, \end{aligned} \tag{4.3}$$

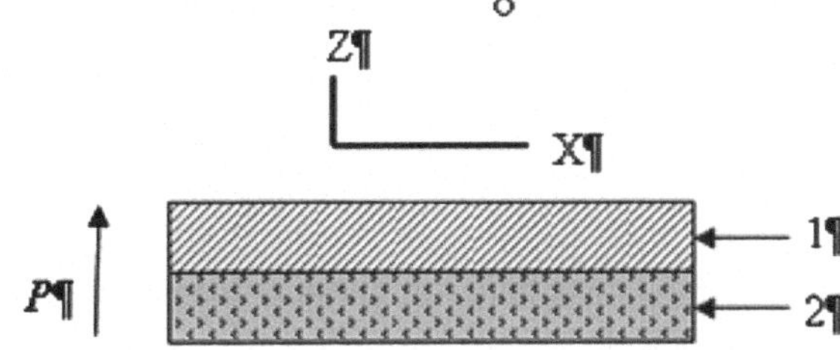

Fig. 4.1 Scheme of bilayer of piezoelectric (1) and piezomagnetic (2) phases. The indicator specifies a direction of polarization

$$ {}^{p}D_3 = {}^{p}\varepsilon_{33}E_3 + {}^{p}d_{31}({}^{p}T_1 + {}^{p}T_2), \tag{4.4}$$

where ${}^{p}T_i$ is the stress and ${}^{p}s_{ii}$ is the compliance of the ferroelectric at constant electric field, ${}^{m}T_i$ is the stress and ${}^{m}s_{ii}$ is the compliance of the ferromagnetic at constant magnetic field, ${}^{p}\varepsilon_{33}$ is the relevant component of the electrical permittivity, ${}^{p}d_{31}$ is the piezoelectric coefficient of the ferroelectric, ${}^{m}q_{ij}$ is the piezomagnetic coefficient of ferromagnetic, E_3 and H_3 are ac electric and magnetic fields. Close to EMR we can assume ${}^{m}T_1 >> {}^{m}T_2$ and ${}^{p}T_1 >> {}^{p}T_2$ (axis 1 is directed along the plate length) such that ${}^{m}T_2$ and ${}^{p}T_2$ may be ignored.

Expressions for the stress components ${}^{p}T_1$ and ${}^{m}T_1$ can be found from 4.2 and 4.3. Substituting these expressions into 4.1 yields a differential equation for u_x. Assuming harmonic motion along x, we get the solution of this equation in the form:

$$u_x = \omega A \cos(\omega x) + B \sin(\omega x), \tag{4.5}$$

where $k = \omega\sqrt{\bar{\rho}\left[\frac{V}{{}^{p}s_{11}} + \frac{1-V}{{}^{m}s_{11}}\right]^{-1}}$ and ω is angular frequency. The integration constants A and B can be found from the boundary conditions. Assuming that the sample surfaces at $x = 0$ and $x = L$ are free from external stresses, we have the following boundary conditions

$$V^{p}T_1 + (1 - V)^{m}T_1 = 0, \text{ at } x = 0 \text{ and } x = L, \tag{4.6}$$

For determining the ME voltage coefficient, we use the open circuit condition:

$$\int_0^L {}^{p}D_3 dx = 0 \tag{4.7}$$

Substituting 3.4 into 3.7 and taking into account 3.5 and 3.6, we can derive

$$\alpha_{E,33} = \frac{2^{p}d_{31}{}^{m}q_{31}{}^{p}s_{11}V(1-V)\tan(kL/2)}{s_1({}^{p}d_{31}^2 - {}^{p}s_{11}{}^{p}\varepsilon_{33})kL - 2^{p}d_{31}^2 V^{m}s_{11}\tan(kL/2)}, \tag{4.8}$$

where $s_1 = V^{m}s_{11} + (1 - V)^{p}s_{11}$.

For transverse field orientation, magnetic induction B has the only component B_1 which should obey the condition $\partial B_1/\partial x = 0$ since B is divergence free. For this case, 3.3 should be written in the more convenient form:

$$ {}^{m}S_1 = {}^{m}s_{11}^{B}\, {}^{m}T_1 + {}^{m}g_{11}B_1, \tag{4.9} $$

where ${}^{m}s_{11}^{B}$ is compliance at constant magnetic induction and ${}^{m}g_{11}$ is piezomagnetic coefficient, ${}^{m}g_{11} = \partial^{m}S_1/\partial B_1$.

In a similar manner as the calculation above, the transverse ME voltage coefficient can shown to be

$$ \alpha_{E,31} = \frac{2^{p}d_{31}{}^{m}g_{11}\mu_{eff}{}^{p}s_{11}V(1-V)\tan(kL/2)}{s_2({}^{p}d_{31}^{2} - {}^{p}s_{11}{}^{p}\varepsilon_{33})kL - 2^{p}d_{31}^{2}V^{m}s_{11}^{B}\tan(kL/2)}, \tag{4.10} $$

Where $s_2 = V^{m}s_{11}^{B} + (1-V)^{p}s_{11}$ Effective permeability μ_{eff} can be found from constitutive equation

$$ H_1 = -{}^{m}g_{11}^{m}T_1 + B_1/{}^{m}\mu_{11}, \tag{4.11} $$

where ${}^{m}\mu_{11}$ is permeability of magnetic phase.

Expressing ${}^{m}T_1$ from 3.3 and substituting it into 4.11 enables finding μ_{eff} with the use of 3.5 after integrating over the sample length.

$$ \mu_{\text{eff}} = \frac{s_2{}^{m}s_{11}^{B}\,{}^{m}\mu_{11}kL}{({}^{m}s_{11}^{B} + {}^{m}g_{11}^{2}\,{}^{m}\mu_{11})kLs_2 + 2^{m}g_{11}^{2}\,{}^{m}\mu_{11}{}^{p}s_{11}(1-V)\tan(kL/2)}. \tag{4.12} $$

As one can see from 4.8 and 4.10 that the value of the ME coefficient under applied fields is directly proportional to the product of piezoelectric d_{31} and piezomagnetic q_{31} or g_{11} modules. Bear in mind, that in reality, there are always loss factors that must be included, even in "perfect" materials if for no other reason than losses associated with electrical contacts. Said loss factors define the width of the resonant line, limiting the peak value of the ME coefficient. The width of the resonant peak can be varied through attenuation coefficients. Such coefficients are present in k and ω, as they are both complex parameters. We shall use a complex frequency $\omega(1 + i/Q)$ with Q to be determined experimentally.

The roots of the denominator in 4.8 and 4.10 define the maxima in the frequency dependence of the ME voltage coefficient. In Figs. 4.2 and 4.3, the frequency dependence of the transverse and longitudinal ME voltage coefficients is shown for the bilayer of nickel ferrite (NFO) and lead zirconate titanate (PZT). In these figures, the resonance peaks caused by oscillations along the x-axes can be seen. The maximum value of the ME coefficient (5400 mV/(cmOe)) is observed for the transverse field orientation, whereas the value at frequency of 100 Hz is 144 mV/(cmOe). Thus, the resonant value of the ME coefficient exceeds the low-frequency value by a factor of about 30. For the longitudinal field orientation, the

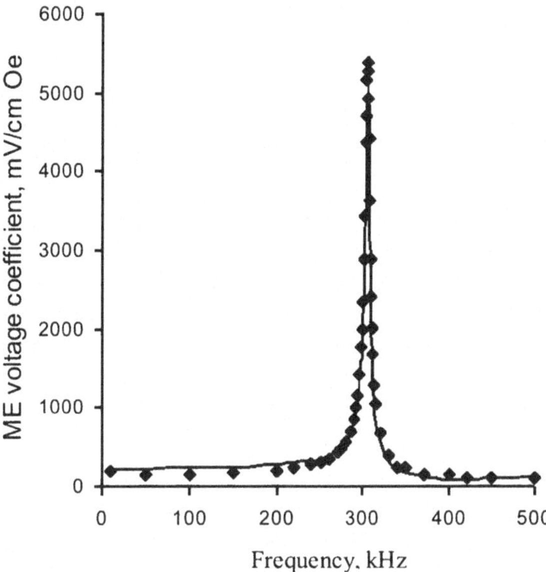

Fig. 4.2 Frequency dependence of transverse ME voltage coefficient for the bilayer of NFO and PZT of 7.3 mm length. Q = 250 and PZT volume fraction is 0.6. There is a good agreement between calculation (*solid line*) and data (*points*)

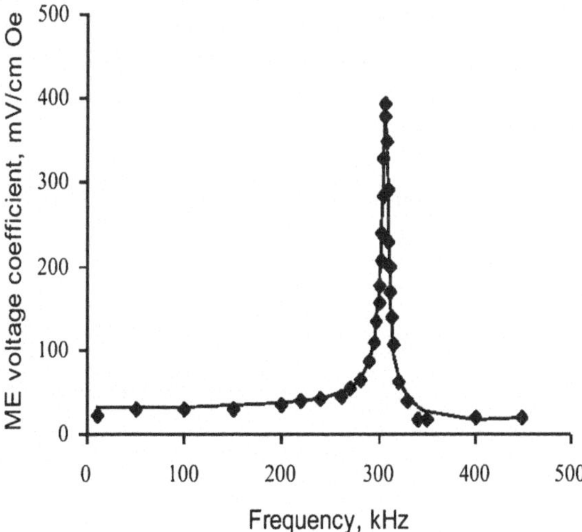

Fig. 4.3 Frequency dependence of longitudinal ME voltage coefficient for the bilayer of NFO and PZT of 7.3 mm length. Q = 105 and PZT volume fraction is 0.6. There is a good agreement between calculation (*solid line*) and data (*points*)

magnitude of the ME effect is smaller by one order of magnitude. This is explained by the fact that, for the longitudinal field orientation, the ME effect is significantly affected by demagnetizing fields.

4.1.2 Disc-Shaped Bilayer

Let us consider now a ferrite–piezoelectric disk-shaped composite of radius R and thickness d, which has thin metal electrodes deposited on bottom and top surfaces, as shown in Fig. 4.4 (Bichurin et al. 2003, 2010, 2012a, b; Bichurin and Petrov 2010, 2012; Petrov et al. 2009a, b).

Assume that the sample is poled normal to the plane of the electrodes, along the z-axis. DC and ac magnetic fields can be directed either along the normal or in the plane of the contacts, which distinguishes the cases of longitudinal and transverse field orientations, respectively. Due to magnetostriction, application of an ac magnetic field will excite both thickness and radial oscillations. In what follows we consider the low-frequency radial oscillations. Radial oscillations will have a notably lower frequency than thickness mode ones, simple because of the geometrical size limitations of the thickness mode.

The disc is supposed to be thin (i.e., $d << R$). Since the surfaces are free from external forces, the normal components of the stress tensors will be zero. For a thin disc, it is reasonable to assume that the component of the stress tensor T_3 is zero not only on the surfaces, but also in the volume of the disc. In addition, since the top and bottom bases of the disc are equipotential surfaces, only the z component of the electric field vector is non-zero. Then, the constitutive equations are determined by 4.2–4.4 for longitudinal field orientation.

It is convenient to take advantage of the symmetry of a disk by use of a cylindrical coordinate system (z, r, and θ). In this case, the strain and stress tensor components need to be transformed in by the following relations:

$$^{i}S_1 = {}^{i}S_{rr}\cos^2(\theta) - 2^{i}S_{r\theta}\sin(\theta)\cos(\theta) + {}^{i}S_{\theta\theta}\sin^2(\theta), \tag{4.13}$$

$$^{i}S_2 = {}^{i}S_{rr}\sin^2(\theta) + 2^{i}S_{r\theta}\sin(\theta)\cos(\theta) + {}^{i}S_{\theta\theta}\cos^2(\theta), \tag{4.14}$$

$$^{i}T_{rr} = {}^{i}T_1\cos^2(\theta) + 2^{i}T_6\sin(\theta)\cos(\theta) + {}^{i}T_2\sin^2(\theta), \tag{4.15}$$

$$^{i}T_{\theta\theta} = {}^{i}T_1\sin^2(\theta) - 2^{i}T_6\sin(\theta)\cos(\theta) + {}^{i}T_2\cos^2(\theta), \tag{4.16}$$

$$^{i}T_{r\theta} = \sin(\theta)\cos(\theta)\left[{}^{i}T_2 - {}^{i}T_1\right] + \left[\cos^2(\theta) - \sin^2(\theta)\right]{}^{i}T_6, \tag{4.17}$$

where $i = p$ or m. The components of the strain tensor in cylindrical coordinates are defined via the vector displacement u, as follows:

$$^{i}S_{rr} = \partial u_r/\partial\, r, \tag{4.18}$$

$$^{i}S_{\theta\theta} = u_r/r, \tag{4.19}$$

$$^{i}S_{r\theta} = (1/r)\partial\, u_r/\partial\, \theta. \tag{4.20}$$

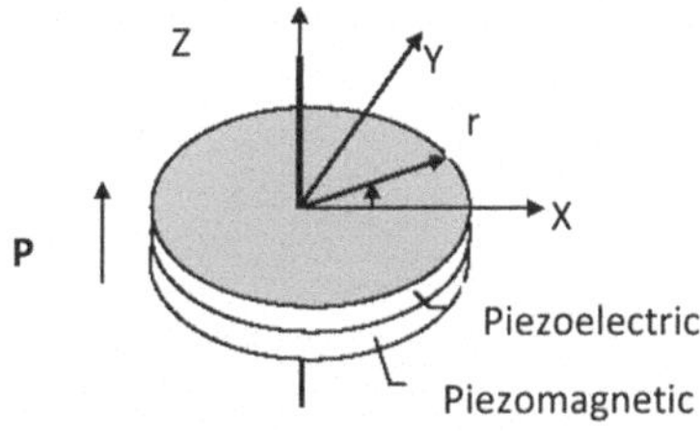

Fig. 4.4 Disk-shaped bilayer of piezoelectric and piezomagnetic phases

The other necessary equation is the equation of medium motion transformed into cylindrical coordinates:

$$-\bar{\rho}\omega^2 u_r = V\left(\frac{\partial^p T_{rr}}{\partial r} + \frac{\partial^p T_{r\theta}}{r\partial\theta} + \frac{^pT_{rr} - {^pT_{\theta\vartheta}}}{r}\right) + (1-V)\left(\frac{\partial^m T_{rr}}{\partial r} + \frac{\partial^m T_{r\theta}}{r\partial\theta} + \frac{^mT_{rr} - {^mT_{\theta\vartheta}}}{r}\right). \tag{4.21}$$

Solving the elasticity equations for stress components and substituting the found expressions into 4.21, we can obtain the equation for the radial displacements. However, the form of this equation depends on the electric and magnetic fields orientations. Next, we consider the ME coupling for longitudinal and transverse fields orientations separately.

4.1.3 Longitudinal Orientation of Electric and Magnetic Fields

In this case, the longitudinal orientation of dc and ac magnetic fields coincides with the direction of polarization. Taking into account the axial symmetry of disk-shaped sample, one can get the following expressions for nonvanishing strain components:

$$\begin{aligned} {^pS_{rr}} &= {^ps_{11}}{^pT_{rr}} + {^ps_{12}}{^pT_{\theta\theta}} + {^pd_{31}}E_3, \\ {^mS_{rr}} &= {^ms_{11}}{^mT_{rr}} + {^ms_{12}}{^mT_{\theta\theta}} + {^mq_{31}}H_3, \\ {^pS_{\theta\theta}} &= {^ps_{12}}{^pT_{rr}} + {^ps_{11}}{^pT_{\theta\theta}} + {^pd_{31}}E_3, \\ {^mS_{\theta\theta}} &= {^ms_{12}}{^mT_{rr}} + {^ms_{11}}{^mT_{\theta\theta}} + {^mq_{31}}H_3 \end{aligned}. \tag{4.22}$$

Solving 3.40 for stress components and substituting these expressions into 3.39, we get the equation of motion in the form:

$$-\omega^2\bar{\rho}u_r = \left[\frac{V}{^ps_{11}(1-{^p\nu^2})} + \frac{1-V}{^ms_{11}(1-{^m\nu^2})}\right]\left(\frac{\partial^2 u_r}{\partial r^2} + \frac{\partial u_r}{r\partial r} - \frac{u_r}{r^2}\right) \tag{4.23}$$

where ${^p\nu}$ and ${^m\nu}$ are Poisson's ratios for piezoelectric and magnetostrictive phases.

The solution of 4.23 is a linear combination of Bessel function of the first and second sort:

$$u_r = c_1 J_1(kr) + c_2 Y_1(kr), \tag{4.24}$$

$$k = \omega\sqrt{\bar{\rho}\left[\frac{V}{{}^p s_{11}(1 - {}^p v^2)} + \frac{1 - V}{{}^m s_{11}(1 - {}^m v^2)}\right]^{-1}}.$$

The constants of integration c_1 and c_2 should be found from boundary conditions: $u_r = 0$ at $r = 0$ and $T_{rr} = 0$ at $r = R$.

For obtaining the ME voltage coefficient, we use the open-circuit condition

$$\int_0^R r dr \int_0^{2\pi} d\theta D_3 = 0 \tag{4.25}$$

where electric induction is determined by the constitutive equation

$$D_3 = {}^p d_{31}({}^p T_{rr} + {}^p T_{\theta\theta}) + {}^p \varepsilon_{33} E_3. \tag{4.26}$$

Substituting 4.22 into 4.24 and 4.25 and then into 3.40, we find the expressions for stress components. Once the stress components are determined, the ME voltage coefficient can be found from 4.26.

$$\alpha_{E,L} = -\frac{2(1 + v)(1 - V)^p s_{11} J_1(kR)^p d_{31}{}^m q_{31}}{(1 - v)^p s_{11}[aJ_0(kR) - (1 - v)s_1 J_1^*]^p \varepsilon_{33} + 2[aJ_0(kR) - s_3 J_1^*]^p d_{31}^2}. \tag{4.27}$$

where $J_1^* = J_1(kR)$, $a = kRs_1, s_1 = V^m s_{11} + (1 - V)^p s_{11}, s_3 = (1 - v)(1 - V)^p s_{11} + 2V^m s_{11}$. It should be noted that ${}^m v$ is assumed to equal ${}^p v = v$ for simplicity of (4.27).

4.1.4 Transverse Orientation of Electric and Magnetic Fields

In this case, the in-plane dc and ac magnetic field vectors are perpendicular to the electric field vector field which is along the z-axis. Equations 3.40 should be replaced by

$$\begin{aligned}
{}^{p}S_{rr} &= {}^{p}s_{11}{}^{p}T_{rr} + {}^{p}s_{12}{}^{p}T_{\theta\theta} + {}^{p}d_{31}E_3,\\
{}^{m}S_{rr} &= {}^{m}s_{11}^{B}{}^{m}T_{rr} + {}^{m}s_{12}^{B}{}^{m}T_{\theta\theta} + [{}^{m}g_{11}\cos^2(\theta) + {}^{m}g_{12}\sin^2(\theta)]B_1,\\
{}^{p}S_{\theta\theta} &= {}^{p}s_{12}{}^{p}T_{rr} + {}^{p}s_{11}{}^{p}T_{\theta\theta} + {}^{p}d_{31}E_3,\\
{}^{m}S_{\theta\theta} &= {}^{m}s_{12}^{B}{}^{m}T_{rr} + {}^{m}s_{11}^{B}{}^{m}T_{\theta\theta} + [{}^{m}g_{12}\cos^2(\theta) + {}^{m}g_{11}\sin^2(\theta)]B_1,\\
{}^{p}S_{r\theta} &= ({}^{p}s_{11} - {}^{p}s_{12}){}^{p}T_{r\theta},\\
{}^{m}S_{r\theta} &= ({}^{m}s_{11}^{B} - {}^{m}s_{12}^{B}){}^{m}T_{r\theta} - {}^{1}\!/\!_{2}\sin(2\theta)({}^{m}g_{11} - {}^{m}g_{12})B_1
\end{aligned} \tag{4.28}$$

Solving 4.28 for stress components and substituting these expressions into 4.25 results in the following form of equation of media motion:

$$\begin{aligned}
-\omega^2\bar{\rho}u_r = &\left[\frac{V}{{}^{p}s_{11}(1-{}^{p}v^2)} + \frac{1-V}{{}^{m}s_{11}(1-{}^{m}v^2)}\right]\left(\frac{\partial^2 u_r}{\partial r^2} + \frac{\partial u_r}{r\partial r} - \frac{u_r}{r^2}\right)\\
&+ \left[\frac{V}{{}^{p}s_{11}(1+{}^{p}v)} + \frac{1-V}{{}^{m}s_{11}(1+{}^{m}v)}\right]\frac{\partial^2 u_r}{r^2\partial\theta^2}
\end{aligned} \tag{4.29}$$

The solution of 4.29 corresponding to the radial mode is defined by 4.24 with $k = \omega\sqrt{\bar{\rho}\left[\frac{V}{{}^{p}s_{11}(1-{}^{p}v^2)} + \frac{1-V}{{}^{m}s_{11}^{B}(1-{}^{m}v^2)}\right]^{-1}}$. The constants of integration c_1 and c_2 should be found from boundary conditions: $u_r = 0$ at $r = 0$ and $\int_0^{2\pi} T_{rr}d\theta = 0$ at $r = R$.

The ME voltage coefficient can be found as

$$\alpha_{E,T} = \frac{(1+v)(1-V){}^{p}s_{11}J_1(kR){}^{p}d_{31}({}^{m}g_{11} + {}^{m}g_{12})\mu_{eff}}{(1-v){}^{p}s_{11}[aJ_0(kR) - (1-v)s_1J_1^*]{}^{p}\varepsilon_{33} + 2[aJ_0(kR) - s_3J_1^*]{}^{p}d_{31}^2} \tag{4.30}$$

where $a = kRs_1$, $s_1 = V{}^{m}s_{11}^{B} + (1-V){}^{p}s_{11}$, $s_3 = (1-v)(1-V){}^{p}s_{11} + 2V{}^{m}s_{11}^{B}$ Effective permeability μ_{eff} can be found similarly to Sect. 3.1. It should be noted that ${}^{m}v$ is assumed to equal ${}^{p}v$ for simplicity of (4.30).

Experimental investigations of the ME effect have been performed for bilayer of NFO and PZT. The samples were discs with the radius of $R = 4.7$ mm, and were poled by dc electric field $E = 4$ kV/mm for 3 h at 80 °C. The bias field dependence of low-frequency ME effect was measured in the beginning. Measurements were done for both longitudinal (see Fig. 4.5) and transverse (see Fig. 4.6) orientations of electric and magnetic fields in the EMR region. Experimental data and corresponding theoretical estimates based on 4.30 are given in Figs. 4.5 and 4.6, for longitudinal and transverse fields respectively. Calculations were performed using the following values of the materials parameters: for NFO ${}^{m}s_{11} = 6.5\cdot10^{-12}$ m²/м, ${}^{m}s_{12} = -2.4\cdot10^{-12}$ m²/м, ${}^{m}q_{31} = 70\cdot10^{-12}$ м/A, ${}^{m}q_{11} = -430\cdot10^{-12}$ м/A, ${}^{m}q_{12} = 125\cdot10^{-12}$ м/A, ${}^{m}\varepsilon_{33}/\varepsilon_0 = 10$; and for PZT ${}^{p}s_{11} = 15.3\cdot10^{-12}$ m²/м, ${}^{p}s_{12} = 5\cdot10^{-12}$ m²/м, ${}^{p}d_{31} = -175\cdot10^{-12}$ м/B, ${}^{p}\varepsilon_{33}/\varepsilon_0 = 1750$. The attenuation parameter was empirically defined by the line-width of the EMR peak at its half-maximum point.

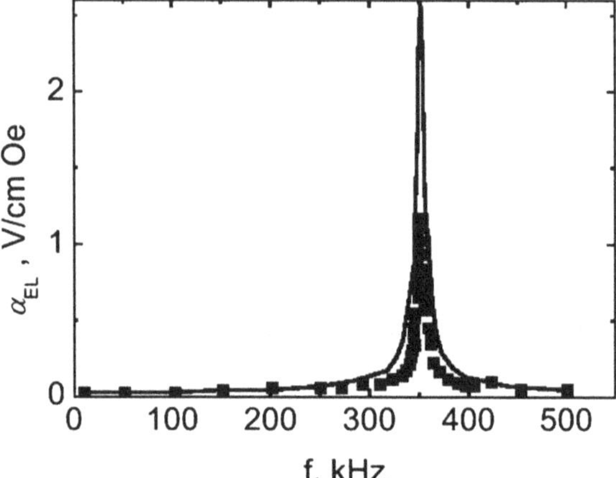

Fig. 4.5 Frequency dependence of magnetoelectric voltage coefficient at longitudinal orientation of fields. *Solid line*—theory, points—experiment

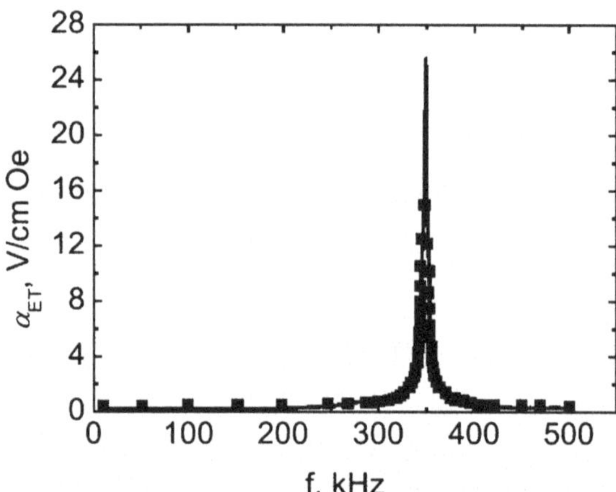

Fig. 4.6 Frequency dependence of magnetoelectric voltage coefficient at transverse orientation of fields. *Solid line*—theory, points—experiment

As can be seen in these figures, there is a relatively good agreement between theory and experimental data. The maximum value of the transverse ME coefficient for the disc-shaped composites was about 15 V/cmOe at resonance, whereas the low-frequency value was only about 0.16 V/cmOe. The coefficient of attenuation was found to be smaller for transverse fields compared to longitudinal ones. This can be accounted for by higher energy losses for longitudinal fields orientation due to Eddy currents in the electrodes. In general, it was found that the transverse ME coefficient was larger than the longitudinal one. As mentioned earlier, this is also due to the fact that the demagnetization field appears at longitudinal orientations that reduces the piezomagnetic modulus.

4.2 Bending Modes

A key drawback for ME effect at longitudinal modes is that the frequencies are quite high, on the order of hundreds of kHz, for nominal sample dimensions. The eddy current losses for the magnetostrictive phase can be quite high at such frequencies, in particular for transition metals and alloys and earth rare alloys such as Terfenol-D, resulting in an inefficient magnetoelectric energy conversion. In order to reduce the operating frequency, one must therefore increase the laminate size that is inconvenient for any applications. An alternative for getting a strong ME coupling is the resonance enhancement at bending modes of the composite. The frequency of applied ac field is expected to be much lower compared to longitudinal acoustic modes. Recent investigations have showed a giant ME effect at bending modes in several layered structures (Petrov et al. 2009a, b, 2012; Fetisov et al. 2011; Sreenivasulu et al. 2011, 2012; Mandal et al. 2010; Laletin and Petrov 2011; Bichurin and Petrov 2013). In this section, we focus our attention on theoretical modeling of ME effects at bending modes.

An in-plane bias field is assumed to be applied to magnetostrictive component to avoid the demagnetizing field. The thickness of the plate is assumed to be small compared to remaining dimensions. Moreover, the plate width is assumed small compared to its length. In that case, we can consider only one component of strain and stress tensors in the EMR region. The equation of bending motion of bilayer has the form:

$$\nabla^2\nabla^2 w + \frac{\rho\, b}{D}\frac{\partial^2 w}{\partial \tau^2} = 0 \tag{4.31}$$

where $\nabla^2\nabla^2$ is biharmonic operator, w is the deflection (displacement in z-direction), t and ρ are thickness and average density of sample, $b = {}^p t + {}^m t$, $\rho = ({}^p\rho^p t + {}^m\rho^m t)/b, {}^p\rho, {}^m\rho$, and ${}^p t, {}^m t$, are densities and thicknesses of piezoelectric and piezomagnetic, correspondingly, and D is cylindrical stiffness.

The boundary conditions for $x = 0$ and $x = L$ have to be used for finding the solution of above equation. Here L is length of bilayer. As an example, we consider the plate with free ends. At free end, the turning moment M_1 and transverse force V_1 equal zero: $M_1 = 0$ and $V_1 = 0$ at $x = 0$ and $x = L$, where $M_1 = \int_A zT_1\, dz_1$, $V_1 = \frac{\partial M_1}{\partial x}$, and A is the cross-sectional area of the sample normal to the x-axis. We are interested in the dynamic ME effect; for an ac magnetic field H applied to a biased sample, one measures the average induced electric field and calculates the ME voltage coefficient. Using the open circuit condition, the ME voltage coefficient can be found as

$$\alpha_{E\,31} = \frac{E_3}{H_1} = -\frac{\int\limits_{z_0-{}^p t}^{z_0} {}^p E_3 dz}{t\, H_1 {}^p\varepsilon_{33}} \tag{4.32}$$

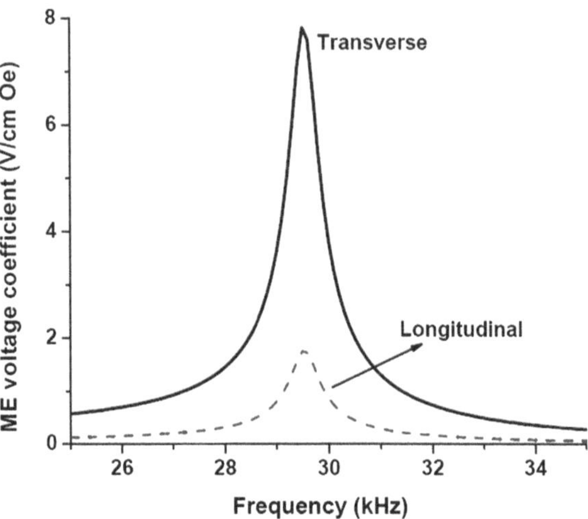

Fig. 4.7 Frequency dependence of longitudinal and transverse ME voltage coefficients for a bilayer of permendur and PZT showing the resonance enhancement of ME interactions at the bending mode frequency. The bilayer is free to bend at both ends. The sample dimensions are $L = 9.2$ mm and total thickness $t = 0.7$ mm and the PZT volume fraction $v = 0.6$

where E_3 and H_1 are the average electric field induced across the sample and applied magnetic field.

Solving the elasticity equations for stress components taking into account 4.31 and substituting the solutions into 4.32 enables one to obtain the explicit expression for ME voltage coefficient

$$\alpha_{E31} = \frac{A \cdot {}^{p}d_{31} \cdot {}^{p}Y \cdot {}^{m}q_{11}}{\Delta \cdot {}^{p}\varepsilon_{33}} F(kL), \tag{4.33}$$

where $A = \frac{{}^{m}Y \cdot {}^{m}t \cdot (2 \cdot z_0 + {}^{m}t) \cdot (2 \cdot z_0 - {}^{p}t)}{4 \cdot D \cdot (1 - {}^{p}K_{31}^2)}$, D is cylindrical stiffness of sample, $k^4 = \omega^2 \rho t / D$, L is the sample length, $F(kL) = (4 \cdot r_2 \cdot r_3 - 2 \cdot r_2 \cdot r_3^2 + 2 \cdot r_2^2 \cdot r_4 \;\; -2 \cdot r_2 \cdot r_4^2 + 4 \cdot r_1 \cdot r_4 - 2 \cdot r_4 \cdot r_1^2 - 2 \cdot r_4 - 2 \cdot r_2) \cdot a_1$; $\Delta = (r_3^2 - r_2^2 - 2 \cdot r_1 \cdot r_3 + r_1^2 + r_4^2)kL$ $+(4 \cdot r_2 \cdot r_3 - 2 \cdot r_2 \cdot r_3^2 + 2 \cdot r_2^2 \cdot r_4 - 2 \cdot r_2 \cdot r_4^2 + 4 \cdot r_1 \cdot r_4 - 2 \cdot r_4 \;\; \cdot r_1^2 - 2 \cdot r_4 - 2 \cdot r_2) \cdot a_1$; $r_1 = cosh(kL)$, $r_2 = sinh(kL)$, $r_3 = cos(kL)$, $r_4 = sin(kL)$, $a_1 = {}^{p}K_{31}^2 \cdot {}^{p}Y^E \cdot ((z_0 - {}^{p}t)^3 - z_0^3)/[3 \cdot D \cdot (1 - {}^{p}K_{31}^2)] + {}^{m}K_{31}^2 \cdot {}^{m}Y \cdot ((z_0 + \; {}^{m}t)^3 z_0^3)/(3 \cdot D)$, ${}^{p}Y$ and ${}^{m}Y$ are the modulus of elasticity of piezoelectric and piezomagnetic component, $z_0 = \frac{1}{2} \frac{{}^{p}Y \, {}^{p}t^2 \; - {}^{m}Y \, {}^{m}t^2}{{}^{p}Y \, {}^{p}t + {}^{m}Y \, {}^{m}t}$.

The energy losses are taken into account by substituting ω for complex frequency $\omega' + i\omega''$ with $\omega''/\omega' = 10^{-3}$.

As an example, we apply 4.33 to the bilayer of permendur and PZT. Figure 4.7 shows the frequency dependence of ME voltage coefficient at bending mode for free-standing bilayer with length 9.2 mm and thickness 0.7 mm for PZT volume fraction 0.67. Graph of $\alpha_{E,31}$ reveals a giant value $\alpha_{E\ 31} = 6.6\ V/cm\ Oe$ and resonance peak lies in the infralow frequency range. Figure 4.8 reveals the theoretical and measured frequency dependencies of transverse ME voltage coefficients for a permendur-PZT bilayer that is free to bend at both ends.

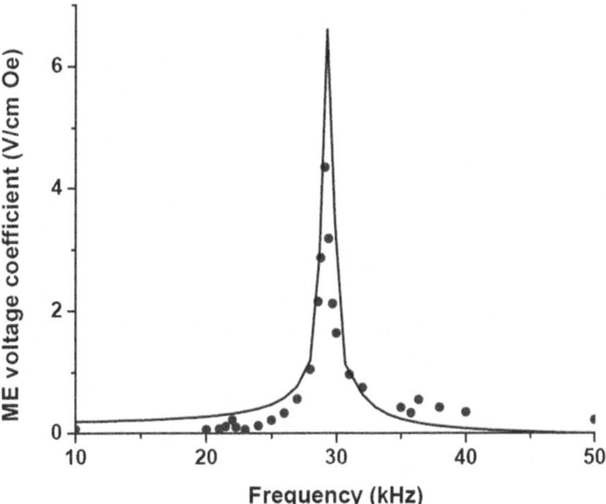

Fig. 4.8 Theoretical (*line*) and measured (*circles*) frequency dependence of transverse ME voltage coefficients for a permendur-PZT bilayer that is free to bend at both ends and with $\nu = 0.67$

According to our model, there is a strong dependence of resonance frequency on boundary conditions. The lowest resonance frequency is expected for the bilayer clamped at one end. One expects bending motion to occur at decreasing frequencies with increasing bilayer length or decreasing thickness.

4.3 Shear Vibrations

Mechanical oscillations of a ME composite can be induced either by alternating magnetic or electric fields. If the length of the electromagnetic wave exceeds the spatial size of the composite by some orders of magnitude, then it is possible to neglect gradients of the electric and magnetic fields within the sample volume. Therefore, based on elastodynamics and electrostatics, the equation of shear medium motion of piezoelectric phase is governed by

$$\rho_p \frac{\partial^2 u_1}{\partial t^2} = \frac{\partial T_{p5}}{\partial z}, \tag{4.34}$$

where u_i is the displacement vector component, ρ_p is the mass density of piezoelectric. Similar equations of media shear motion can be written for magnetostrictive phases and substrate (Fig. 4.9). To solve these equations for displacement component, one should use the appropriate boundary conditions for stress and strain components on interfaces. Substituting the found solutions into the open circuit condition enables one to obtain the expression for ME voltage coefficient (Bichurin and Petrov 2013) (Fig. 4.9).

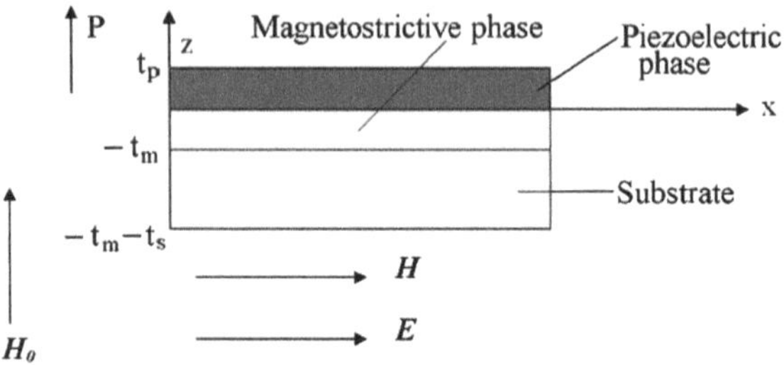

Fig. 4.9 Bilayer of piezoelectric and ferrite in the *x-z* plane on the substrate. The poling electric field *E*0 is along *z* direction and ac electric field *E* is along *x*. The bias magnetic field *H*0 and the alternating magnetic field *H* are along *z* and *x*, respectively

Expressions for the strain in the ferromagnetic layers S_{mi}, the strain in the ferroelectric layers S_{pi}, and strain in the substrate S_{si} have the form:

$$\begin{cases} S_{p5} = d_{15}E_1 + s_{p44}T_{p5}(z) \\ S_{5m} = q_{15}H_1 + s_{m44}T_{m5}(z) \\ S_{s5} = s_{s44}T_{s5}(z) \end{cases} \tag{4.35}$$

where T_{pi} is the stress and s_{pii} is the compliance of the ferroelectric at constant electric field, T_{mi} is the stress and s_{mii} is the compliance of the ferromagnetic at constant magnetic field, d_{ij} is the piezoelectric coefficient of the ferroelectric, q_{ij} is the piezomagnetic coefficient of ferromagnetic, E_1 and H_1 are electric and magnetic fields.

Electric displacements *D* for the piezoelectric phases is given by

$$D_p(z) = d_{15}T_{p5}(z) + \varepsilon_{33}E_1 \tag{4.36}$$

where ε_{33} is the relevant component of the electrical permittivity.

For determining the ME voltage coefficient, we use the open circuit condition $\int_0^{t_p} D_p(z)dz = 0$. Substituting expression (4.35) into (4.34) and solving the open circuit condition for E_1 enables finding the expression for the ME voltage coefficient $\alpha = E_1/H_1$ for the free standing bilayer:

$$\begin{aligned} \alpha_E = 2k_p{}^{d_{15}}/_{\varepsilon_{33}}q_{15}[1-(1-Q_4)Q_2-Q_4]\{[(2s_{m44}Q_4+s_{p44}t_pk_mQ_3)k_pQ_2 \\ +(k_p^2t_ps_{m44}Q_4-s_{p44}k_mQ_3)Q_1-2s_{m44}k_pQ_4]K_p^2-s_{p44}k_pt_pk_mQ_5-k_p^2t_ps_{m44}Q_6\}^{-1}, \end{aligned} \tag{4.37}$$

where $k_m = \omega\sqrt{\rho_m s_{m44}}$, $k_p = \omega\sqrt{\rho_p s_{p44}}$, $Q_5 = Q_2Q_3$, $Q_6 = Q_1Q_4$, $Q_1 = \sin(k_pt_p)$, $Q_2 = \cos(k_pt_p)$, $Q_3 = \sin(k_mt_m)$, $Q_4 = \cos(k_mt_m)$,ω is the circular frequency.

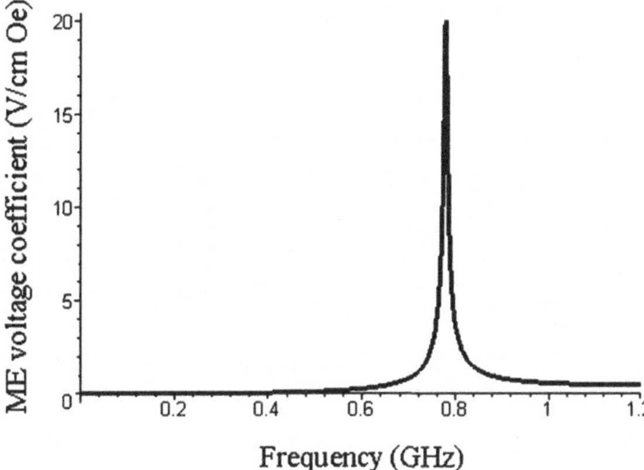

Fig. 4.10 Frequency dependence of ME voltage coefficient for a free bilayer structure of PZT of thickness 300 nm and NFO of thickness 2 [1]m and without substrate

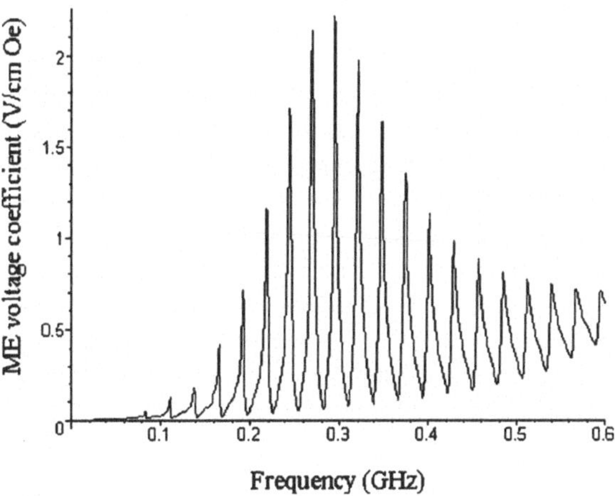

Fig. 4.11 Frequency dependence of ME voltage coefficient for bilayer of 300 nm thick PZT and 2 μm thick NFO on 100 μm thick SrTiO3 substrate

As an example, Fig. 4.10 shows a resonances peak at shear modes in free standing bilayer of lead zirconate titanate (PZT) and nickel ferrite $NiFe_2O_4$. The maximum value of the ME coefficient (20 V/(cm Oe)) is observed at frequency of about 0.8 GHz. Placing the magnetostrictive-piezoelectric film on a substrate is expected to lead to two significant effects: a decrease in the EMR frequency due to clamping and a fine structure in the ME voltage versus frequency spectrum. Figure 4.11 shows a family of equally spaced peaks that are linear combinations of the modes corresponding to the substrate and bilayer structure. The frequency separation between two consecutive peaks is determined primarily by the substrate thickness. The envelope of the fine structure shows a maximum at 0.3 GHz. This maximum is due to EMR in ME bilayer. Peak value of ME voltage coefficient (2.2 V/cm Oe) is a factor of about ten smaller than for free standing bilayer.

4.4 Conclusions

In this chapter, we have presented a theory for the resonance enhancement of ME interactions at frequencies corresponding to EMR. Frequency dependence for longitudinal and transverse ME voltage coefficients are obtained using the simultaneous solution of electrostatic, magnetostatic, and elastodynamic equations. The ME voltage coefficients are estimated from known material parameters (piezoelectric coupling, magnetostriction, elastic constants, etc.) of composite components.

It is shown that the ME coupling in the EMR region exceeds the low-frequency value by more than an order of magnitude. It was found that the peak transverse ME coefficient at EMR is larger than the longitudinal one. This is accounted for (i) by higher energy losses for longitudinal fields orientation due to eddy currents in the electrodes and (ii) by influence of demagnetization field that appears at longitudinal orientations that reduces the piezomagnetic modulus.

A theory has been described for ME interactions at shear modes in a magnetostrictive-piezoelectric film on a substrate in EMR region. For a NFO-PZT structure on SrTiO3 substrate, the strength of ME interactions is weaker than for thick film bilayers due to the strong clamping effects of the substrate. For increasing substrate thickness in a bilayer, the ME coefficient drops rapidly and the EMR frequency decreases. For a magnetostrictive-piezoelectric film on a substrate, the ME voltage versus frequency profile shows a fine structure consisting of equally spaced peaks. The distance between the peaks is a function of the substrate thickness. The obtained phenomenon is of importance for the realization of multifunctional ME sensors/transducers operating at microwave frequencies.

The results of calculations obtained for a NFO spinel–PZT composite are in good agreement with the experimental data.

References

Bichurin MI, Filippov DA, Petrov VM, Laletsin VM, Paddubnaya NN, Srinivasan G (2003) Resonance magnetoelectric effects in layered magnetostrictive-piezoelectric composites. Phys Rev B, 68:132408(1)–132408(4)

Bichurin MI Magnetoelectric interaction in magnetically ordered materials (Review) (2012) In: Bichurin MI, Viehland D (eds) Magnetoelectricity in composites. Pan Stanford Publishing, Singapore, pp 1–24

Bichurin MI, Petrov VM () Magnetoelectric effect in the electromechanical resonance range (2012) In: Bichurin MI, Viehland D (eds) Magnetoelectricity in composites. Pan Stanford Publishing, Singapore, pp 91–104

Bichurin MI, Petrov VM (2010) Magnetoelectric effect in magnetostriction-piezoelectric multiferroics. Low Temp Phys 36:544–549

Bichurin MI, Petrov VM (2012) Modeling of magnetoelectric interaction in magnetostrictive-piezoelectric composites. Adv Condens Matter Phys 2012:798310

Bichurin MI, Petrov VM, Averkin SV, Filippov AV (2010b) Electromechanical resonance in magnetoelectric layered structures. Phys Solid State 52:2116

Bichurin MI, Petrov VM, Averkin SV, Liverts E (2010a) Present status of theoretical modeling the magnetoelectric effect in magnetostrictive-piezoelectric nanostructures. Part I: low frequency and electromechanical resonance ranges. J Appl Phys 107:053904

Bichurin MI, Petrov VM, Petrov RV (2012a) Direct and inverse magnetoelectric effect in layered composites in electromechanical resonance range: a review. J Magn Magn Mater 324:3548–3550

Bichurin MI, Petrov RV, Petrov VM (2013) Magnetoelectric effect at thickness shear mode in ferrite-piezoelectric bilayer. Appl Phys Lett 103:0929021

Bichurin MI, Petrov VM, Petrov RV, Priya S (2012b) Electromechanical resonance in magnetoelectric composites: direct and inverse effect. Solid State Phenom 189:129–143

Fetisov LY, Perov NS, Fetisov YK, Srinivasan G, Petrov VM (2011) Resonance magnetoelectric interactions in an asymmetric ferromagnetic-ferroelectric layered structure. J Appl Phys 109:053908

Filippov DA, Bichurin MI, Petrov VM, Laletin VM, Poddubnaya NN, Srinivasan G (2004) Giant magnetoelectric effect in composite materials in the region of electromechanical resonance. Tech Phys Lett 30(1):15–20

Laletin VM, Petrov VM (2011) Nonlinear magnetoelectric response of a bulk magnetostrictive–piezoelectric composite. Solid State Commun 151:1806–1809

Mandal SK, Sreenivasulu G, Petrov VM, Srinivasan G (2010) Flexural deformation in a compositionally stepped ferrite and magnetoelectric effects in a composite with piezoelectrics. Appl Phys Lett 96:192502

Petrov VM, Bichurin MI, Srinivasan G, Laletin VM, Petrov RV (2012) Bending modes and magnetoelectric effects in asymmetric ferromagnetic-ferroelectric structure. Solid State Phenom 190:281–284

Petrov VM, Bichurin MI, Zibtsev VV, Mandal SK, Srinivasan G (2009b) Flexural deformation and bending mode of magnetoelectric nanobilayer. J Appl Phys 106:113901

Petrov VM, Srinivasan G, Bichurin MI, Galkina TA (2009a) Theory of magnetoelectric effect for bending modes in magnetostrictive-piezoelectric bilayers. J Appl Phys 105:063911

Sreenivasulu G, Mandal SK, Petrov VM, Mukundan A, Rengesh S (2011) Bending resonance in a magnetostrictive-piezoelectric bilayer and magnetoelectric interactions. Integr Ferroelectr 126:87–93

Sreenivasulu G, Petrov VM, Fetisov LY, Fetisov YK, Srinivasan G (2012) Magnetoelectric interactions in layered composites of piezoelectric quartz and magnetostrictive alloys. Phys Rev B 86:214405

Zhang N, Petrov VM, Johnson T, Mandal SK, Srinivasan G (2009) Enhancement of magnetoelectric coupling in a piezoelectric-magnetostrictive semiring structure. J Appl Phys 106:126101

Chapter 5
Magnetic Resonance in Composites

Abstract In this chapter, we address the electric field induced magnetic resonance field shift in composites of ferrite and piezoelectric components. A phenomenological theory is proposed to treat the ME coupling at frequencies corresponding to ferromagnetic resonance in a multilayer consisting of alternate layers of piezoelectric and magnetostrictive phases. We discuss two models: a simple bimorph structure and a generalized approach in which the multilayer composite is considered as a homogeneous medium. Expressions for the electric field induced magnetic resonance field shift are obtained for both cases. Magnetic resonance field shift is directly proportional to the product of the applied electric field and the ME coupling constant. A method for the calculation of magnetoelectric coefficients from experimental data is presented.

The majority of prior work on magnetoelectric (ME) composites, both theoretical and experimental studies, has been devoted to the low-frequency range (10 Hz–10 kHz). However, ME composites also offer important applications in the microwave range (Nan et al. 2008; Bichurin et al. 1990, 2002a, b, c, d, 2011, 2013; Tatarenko et al. 2006, 2010; Zhai et al. 2007; Lou et al. 2012; Tatarenko and Bichurin 2012; Li et al. 1911; Srinivasan et al. 2009; Petrov et al. 2008). In this frequency range, the ME effect reveals itself as a change in the magnetic permeability under an external electric field. Investigations of a ferromagnetic resonance (FMR) line shift by an applied electric field are easily performed for layered ferrite–piezoelectric structures (Bichurin and Petrov 1988). In addition, layered composites are of interest for applications as electrically tunable microwave phase shifters, devices based on FMR, magnetic-controlled electrooptical and/or piezoelectric devices, and electrically readable magnetic (ME) memories (Bichurin et al. 1990, 2011, 2012; Tatarenko and Bichurin 2012; Li et al. 1911; Srinivasan et al. 2009; Petrov et al. 2007, 2008; Tatarenko et al. 2006; Bichurin and Petrov 1988; Bichurin and Viehland 2012; Srinivasan 2010; Dong et al. 2006; Bichurin and Petrov 2012).

M. Bichurin and V. Petrov, *Modeling of Magnetoelectric Effects in Composites*, Springer Series in Materials Science 201, DOI: 10.1007/978-94-017-9156-4_5,

In this chapter, we discuss theoretical models of multilayer ME composites in the microwave range (Bichurin and Petrov 1994, 1995; Bichurin et al. 2001, 2002). The model is a phenomenological theory, which is based on the magnetic susceptibility as function of strain, whose this strain-dependence is defined both by piezoelectric and the magnetoelastic constants of piezoelectric and magnetostrictive phases, respectively.

5.1 Bilayer Structure

Let us again consider the simple model of a bilayer structure, consisting of a ferrite spinel with cubic (m3 m) symmetry and poled lead zirconate titanate (PZT) with a symmetry of ∞ m. The influence of an electric field on the piezoelectric PZT phase can be described as follows:

$$ {}^{\mathrm{p}}\mathrm{T}_{\mathrm{ij}} = {}^{\mathrm{p}}\mathrm{c}_{\mathrm{ijkl}}\,{}^{\mathrm{p}}\mathrm{S}_{\mathrm{kl}} - \mathrm{e}_{\mathrm{kij}}\mathrm{E}_{\mathrm{k}}; \tag{5.1} $$

where E_k is a component of the electric field vector; and ${}^pT_{ij}$, ${}^pS_{kl}$, e_{kij}, and ${}^pc_{ijk}$ are the components of the stress, strain, piezoelectric, and elastic stiffness tensors of the piezoelectric phase, respectively. The tensor coefficients of the elastic stiffness has the form

$$ {}^pc = \begin{pmatrix} {}^p\tilde{n}_{11} & {}^pc_{12} & {}^pc_{13} & 0 & 0 & 0 \\ {}^pc_{12} & {}^pc_{11} & {}^pc_{13} & 0 & 0 & 0 \\ {}^pc_{13} & {}^pc_{13} & {}^pc_{11} & 0 & 0 & 0 \\ 0 & 0 & 0 & {}^pc_{44} & 0 & 0 \\ 0 & 0 & 0 & 0 & {}^pc_{44} & 0 \\ 0 & 0 & 0 & 0 & 0 & \frac{1}{2}({}^pc_{11} - {}^pc_{12}) \end{pmatrix} \tag{5.2} $$

and that of the piezoelectric coefficients is by

$$ e = \begin{pmatrix} 0 & 0 & 0 & 0 & e_{15} & 0 \\ 0 & 0 & 0 & e_{15} & 0 & 0 \\ e_{31} & e_{31} & e_{33} & 0 & 0 & 0 \end{pmatrix} \tag{5.3} $$

Substituting the 5.2 and 5.3 into 5.1, and by applying an electric field along the axis of polarization (i.e., $E_3 = E$, $E_1 = E_2 = 0$), we obtain the following stresses in the piezoelectric phase:

$$ \begin{aligned} {}^{\mathrm{p}}\mathrm{T}_{33} &= -\,\mathrm{e}_{33}\mathrm{E} + {}^{\mathrm{p}}\mathrm{c}_{13}({}^{\mathrm{p}}\mathrm{S}_{11} + {}^{\mathrm{p}}\mathrm{S}_{22}) + {}^{\mathrm{p}}\mathrm{c}_{33}\,{}^{\mathrm{p}}\mathrm{S}_{33}, \\ {}^{\mathrm{p}}\mathrm{T}_{11} &= -\,\mathrm{e}_{31}\mathrm{E} + {}^{\mathrm{p}}\mathrm{c}_{11}\,{}^{\mathrm{p}}\mathrm{S}_{11} + {}^{\mathrm{p}}\mathrm{c}^{\mathrm{p}}_{12}\mathrm{S}_{22} + {}^{\mathrm{p}}\mathrm{c}_{13}\,{}^{\mathrm{p}}\mathrm{S}_{33} = 0, \\ {}^{\mathrm{p}}\mathrm{T}_{22} &= -\,\mathrm{e}_{31}\mathrm{E} + {}^{\mathrm{p}}\mathrm{c}_{12}\,{}^{\mathrm{p}}\mathrm{S}_{11} + {}^{\mathrm{p}}\mathrm{c}_{11}\,{}^{\mathrm{p}}\mathrm{S}_{22} + {}^{\mathrm{p}}\mathrm{c}_{13}\,{}^{\mathrm{p}}\mathrm{S}_{33} = 0. \end{aligned} \tag{5.4} $$

Now, assume that the axis of polarization in the piezoelectric phase coincides with the crystallographic [111] axis of the magnetostrictive phase. The stresses in the magnetostrictive phase considering are then given by

$$\begin{cases} {}^mT_1 = \left(\frac{{}^mc_{11}+{}^mc_{12}}{2} + {}^mc_{44}\right){}^mS_1 + \frac{{}^mc_{11}+5{}^mc_{12}-2{}^mc_{44}}{6}{}^mS_2 + \frac{{}^mc_{11}+2{}^mc_{12}-2{}^mc_{44}}{3}{}^mS_3 \\ \quad + \frac{{}^mc_{11}-{}^mc_{12}-2{}^mc_{44}}{3\sqrt{2}}{}^mS_4, \\ {}^mT_2 = \frac{{}^mc_{11}+5{}^mc_{12}-2{}^mc_{44}}{6}{}^mS_1 + \left(\frac{{}^mc_{11}+{}^mc_{12}}{2} + {}^mc_{44}\right){}^mS_2 + \frac{{}^mc_{11}+2{}^mc_{12}+4{}^mc_{44}}{3}{}^mS_3, \\ {}^mT_3 = \frac{{}^mc_{11}+2{}^mc_{12}-2{}^mc_{44}}{3}{}^mS_1 + \frac{{}^mc_{11}+2{}^mc_{12}-2{}^mc_{44}}{3}{}^mS_2 + \frac{{}^mc_{11}+2{}^mc_{12}-2{}^mc_{44}}{3}{}^mS_3 \\ \quad - \frac{{}^mc_{11}-{}^mc_{12}-2{}^mc_{44}}{3\sqrt{2}}{}^mS_4, \\ {}^mT_4 = \frac{{}^mc_{11}+2{}^mc_{12}-2{}^mc_{44}}{3}{}^mS_1 - \frac{{}^mc_{11}-{}^mc_{12}-2{}^mc_{44}}{3\sqrt{2}}{}^mS_2 + \frac{{}^mc_{11}-{}^mc_{12}+{}^mc_{44}}{3}{}^mS_4. \end{cases} \tag{5.5}$$

To calculate the ME effect of bilayer structures in the FMR range, we should use the following procedure: (i) ${}^mS_3$ is defined as a function of the stress ${}^mT_3$; (ii) ${}^pT_3$ is defined as a function of the strain ${}^pS_3$; and (iii) use the known expressions for the dependence of the resonant magnetic field on stress to determine the FMR line shift. Thus, the task is reduced to the solutions of electrostatic equations under specific boundary conditions.

Let us then consider that the bilayer structure is mechanically clamped along the 3 axis the ferrite and piezoelectric subsystems. In this case, the boundary conditions, without taking into account forces of friction, have the form

$$\begin{aligned} {}^pT_3 &= {}^mT_3, \\ {}^pS_3 &= -\frac{{}^mv}{{}^pv} \cdot mS_3, \end{aligned} \tag{5.6}$$

where ${}^mv$ and ${}^pv$ are the volume fractions of the magnetostrictive and piezoelectric phases, respectively. From 5.4, 5.5, and 5.6 we can find that the stress along the 3 axis is given by

$${}^mT_3 = \frac{E_3\left(\frac{2{}^pc_{13}e_{31}}{{}^pc_{11}+{}^pc_{12}} - e_{33}\right)}{1 + \frac{{}^mv}{{}^pv}\left(\frac{1}{3{}^mc_{44}} + \frac{1}{3+({}^mc_{11}+2{}^mc_{12})}\right)\left({}^pc_{33} - 2\frac{{}^pc_{13}^2}{{}^pc_{11}+{}^pc_{12}}\right)}. \tag{5.7}$$

It is known that an applied electric voltage results in a change of the resonant magnetic field (Bichurin and Petrov 1995). We limit ourselves to the case when both the magnetic and electric fields are oriented along the polarization axis of the piezoelectric phases, which also corresponds to the [111] axis of the magnetostrictive one. In this case, the shift of the resonant magnetic field is given by

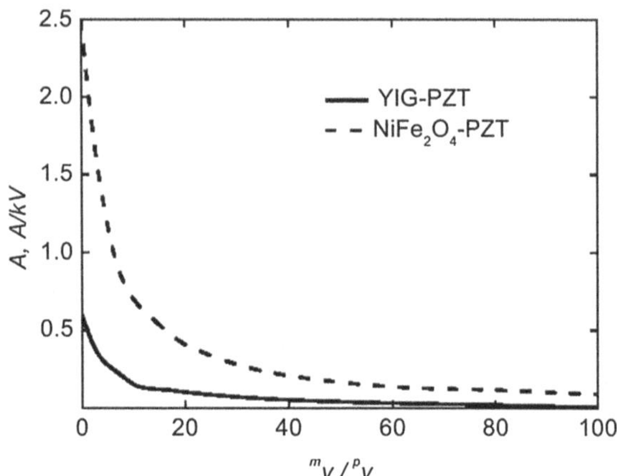

Fig. 5.1 Dependence of magnetoelectric constant A on volume fraction of magnetostrictive and piezoelectric phases ${}^{m}v/{}^{p}v$. The results are for bilayer composites of nickel ferrite–PZT and YIG–PZT

$$\delta H_E = \frac{3\lambda_{111}{}^{m}T_3}{M_0} = AE_3, \tag{5.8}$$

where M_0 is the saturation magnetization of the magnetostrictive layer, and the ME constant is defined by

$$A = \frac{3\lambda_{111}}{M_0} = \frac{\left(\frac{2^{p}c_{13}e_{31}}{{}^{p}c_{11}+{}^{p}c_{12}} - e_{33}\right)}{\left[1 + \frac{{}^{m}v}{{}^{p}v}\left(\frac{1}{3^{m}c_{44}} + \frac{1}{3+({}^{m}c_{11}+2^{m}c_{12})}\right)\left({}^{p}c_{33} - 2\frac{{}^{p}c_{13}^{2}}{{}^{p}c_{11}+{}^{p}c_{12}}\right)\right]}. \tag{5.9}$$

From 5.7 and 5.8, it is possible to estimate the shift of the resonant field for a bilayer magnetostrictive–piezoelectric composite. The theoretical values for the ME constant $A = \delta H_E/E_3$ were determined for composites of both nickel ferrite (NFO)–PZT and yttrium iron garnet (YIG)–PZT. For these calculations, the following materials parameters of two component phases were used: *YIG* ${}^{p}c_{11} = 12.6 \times 10^{10}$ N/m^2, ${}^{p}c_{12} = 7.95 \times 10^{10}$ N/m^2, ${}^{p}c_{13} = 8.4 \times 10^{10}$ N/m^2, ${}^{p}c_{33} = 11.7 \times 10^{10}$ N/m^2, $e_{31} = -6.5$ Sim/m^2, and $e_{33} = 23.3$ Sim/m^2: *$NiFe_2O_4$* ${}^{m}c_{11} = 22 \times 10^{10}$ N/m^2, ${}^{m}c_{12} = 10.9 \times 10^{10}$ N/m^2, ${}^{m}c_{44} = 8.12 \times 10^{10}$ N/m^2, $\lambda_{111} = -21.6 \times 10^{10}$, and $4\pi M_0 = 3200$ Gs.

In Fig. 5.1, the dependence of the ME constant A on the relative volume fraction of the component phases in the bilayer structure is shown. In a composite of NFO and PZT, the ME effect has been reported to be stronger than that of YIG–PZT. This is because NFO has a much higher magnetostriction than YIG. Another notable feature in Fig. 5.1 is the fact that theoretically the field shift should increase with increasing volume fraction of the piezoelectric phase. However, please bear in mind that the magnetic resonance line will become to weak if the volume fraction of the magnetostrictive phase is too low. This is uniquely different than results at lower frequencies, where the largest ME effect is found for an approximate volume fraction ratio of 50:50. Assuming that ${}^{m}v/{}^{p}v = 1$, we can estimate that the resonance line shift under applied electric field is given by the relation $\delta H = \mathrm{AE}$, where $A \approx 2$ Oe-cm/kV for $NiFe_2O_4$–PZT and $A \approx 0.45$ Oe-cm/kV for YIG–PZT.

The values for the calculated resonance line shift are in good agreement with prior experimental data: theoretical line shift is 0.45 Oe-cm/kV, which is comparable to the measured one of 0.2–0.56 Oe-cm/kV depending on relative volume fraction of magnetostrictive and piezoelectric phases. Thus, we can see that it is necessary to use a piezoelectric component with large piezoelectric coefficients, and a magnetostrictive one with small saturation magnetization and high magnetostriction.

5.2 Basic Theory: Macroscopic Homogeneous Model

Now, let us assume that the composite is a macroscopically homogeneous material, i.e., the ME composite can be considered as an effective media with a ME effect that is not observed separately in each component phase. In this case, the influence of an external electric field E on the magnetic resonance spectrum can be described by the term (Bichurin and Petrov 1995; Bichurin et al. 2001, 2002, 2009; Petrov et al. 2008; Shastry et al. 2004)

$$W = \int_V (W_0 + \Delta W_{\mathrm{ME}}) d^3\mathrm{x} \tag{5.10}$$

in the thermodynamic potential; where W_0 is thermodynamic potential at $\mathbf{E} = 0$, and

$$\Delta W_{ME} = B_{ikn} E_i M_k M_n + b_{ijkn} E_i M_k M_n, \tag{5.11}$$

and where B_{ijk} and b_{ijkn} are the linear ME coefficients which are third rank axial tensors. The potential ΔW_{ME} can be found by considering the elastic, magnetostrictive, piezoelectric, and electrostrictive contributions at specified boundary conditions.

First, let us obtain the expression for the magnetic susceptibility tensor of the ferrite phase in the presence of ME interactions. In this case, the composite will be influenced by an applied electric field, and also by the constant and variable magnetic fields that are necessary to observe magnetic resonance. Solving the linearized equation for the magnetization vector rotation, without taking into account losses, and with taking into account the effect of demagnetization factors, the following expression can be obtained (Bichurin et al. 2002) for the magnetic susceptibility tensor:

$$\chi^m = \begin{bmatrix} \chi_1 & \chi_5 - i\chi_a & 0 \\ \chi_5 - i\chi_a & \chi_2 & 0 \\ 0 & 0 & 0 \end{bmatrix}, \quad (5.12)$$

where

$$\chi_1 = D^{-1}\gamma^2 M_0[H_{03'} + M_0 \sum_i (N^i_{1'1'} - N^i_{3'3'})];$$
$$\chi_2 = D^{-1}\gamma^2 M_0[H_{03'} + M_0 \sum_i (N^i_{2'2'} - N^i_{3'3'})];$$
$$\chi_3 = -D^{-1}\gamma^2 M_0^2 \sum_i N^i_{1'2'};$$
$$\chi_a = D^{-1}\gamma M_0 \omega;$$
$$D = \omega_0^2 - \omega^2;$$
$$\omega_0^2 = \gamma^2[H_{03'} + \sum_i (N^i_{1'1'} - N^i_{3'3'})M_0][H_{03'} + \\ + \sum (N^i_{2'2'} - N^i_{3'3'})M_0] - (\sum_i N^i_{1'2'}M_0)^2,$$

and where γ is the magnetomechanical coefficient, ω is circular frequency, $N_{kl}^{i=a}$ are the effective demagnetization factors describing the effective fields of the magnetic anisotropy, $N_{kl}^{i=m}$ is the demagnetization of the sample form; and 1′, 2′, and 3′ is a system of coordinates in which the axis 3′ is directed along that of the spontaneous magnetization.

The following additional term with ($i = E$) must be included into the magnetic anisotropy, to account for the ME constants B_{ij}

$$N^{i'}_{k'n'} = 2B_{ikn}E_{oi}\beta_{k'k}\beta_{n'n} \quad (5.13)$$

where $\hat{\beta}$ is the matrix of direction cosines projected along the axes (1′, 2′, 3′) from the axes of the crystallographic system (1, 2, 3). We must also account for losses in the equation for the magnetization vector rotation, which will take the form of $i\omega\alpha(\boldsymbol{M_0} \times \boldsymbol{m})/M_0$: where α is the loss parameter and m *is* the variable magnetization. This results in a complex value for the magnetic susceptibility, given as $\chi = \chi' + i\,\chi''$ with

$$\chi' = \chi_0 \frac{\omega_0^2(\omega_0^2 - \omega^2 + 2\alpha^2\omega^2)}{(\omega_0^2 - \omega^2)^2 + 4\alpha^2\omega_0^2\omega^2},$$
$$\chi'' = \chi_0 \frac{\alpha\omega\omega_0(\omega_0^2 + \omega^2)}{(\omega_0^2 - \omega^2)^2 + 4\alpha^2\omega_0^2\omega^2}, \chi_0 = \gamma\frac{M_0}{\omega_0}.$$

As an example, we consider a special case when the magnetostrictive component is magnetized in a plane (110) under the corner θ to the cubic axis [001]. Theoretical calculations are essentially simplified by choice of magnetization direction coinciding to direction of stress ${}^{m}T_{33}$. Then the additional term of energy can be presented as

$$\begin{aligned}\Delta W_{\mathrm{ME}} = & \frac{3}{8M^2}(\lambda_{111} - \lambda_{100} + (\lambda_{100} - \lambda_{111})\cos 2\theta)M_1^{2\,m}T_{33} \\ & + \frac{9}{32M^2}(\lambda_{111} - \lambda_{100} + (\lambda_{100} - \lambda_{111})\cos 4\theta)M_2^{2\,m}T_{33} \\ & + \frac{3}{8M^2}\left((\lambda_{100} - \lambda_{111})\sin 2\theta + \frac{3}{2}(\lambda_{100} - \lambda_{111})\sin 4\theta\right)M_2M_3{}^{m}T_{33}\,. \\ & + \frac{3}{8M^2}(\left(\frac{-9\lambda_{111} - 7\lambda_{100}}{4}\right) + (\lambda_{111} - \lambda_{100})\cos 2\theta + \\ & + \frac{3}{4}(\lambda_{111} - \lambda_{100})\cos 4\theta)M_3^{2\,m}T_{33}\end{aligned} \quad (5.14)$$

For the special case considered above when $\boldsymbol{M_0}$//[111], we can obtain

$$\begin{aligned}\Delta W_{\mathrm{ME}} = & \left(\frac{\lambda_{111} - \lambda_{100}}{2}\right)M_1^{2\,m}T_{33} + \left(\frac{\lambda_{111} - \lambda_{100}}{2}\right)M_2^{2\,m}T_{33} \\ & + \left(\frac{-\lambda_{111} - \lambda_{100}}{2}\right)M_3^{2\,m}T_{33}\end{aligned}\,. \quad (5.15)$$

Using 6.12 it is then easy to show that the resonance line is shifted under action of an applied electric field, and using a linear approximation for N_{kl}^E, we have

$$\delta H_E = -\frac{M_0}{Q_1}\left[Q_2(N_{11}^E - N_{33}^E) + Q_3(N_{22}^E - N_{33}^E) + Q_4N_{12}^E\right], \quad (5.16)$$

where

$$Q_1 = 2H_3 + M_0\sum_{i\neq E}[(N_{11}^E - N_{33}^E) + (N_{22}^E - N_{33}^E)];$$

$$Q_2 = \left[H_3 + M_0\sum_{i\neq E}(N_{22}^E - N_{33}^E)\right];$$

$$Q_3 = \left[H_3 + M_0\sum_{i\neq E}(N_{11}^E - N_{33}^E)\right];$$

$$Q_4 = 2M_0\sum_{i\neq E}N_{12}^i.$$

Equation 5.15 allows us to define the ME constants of the composite, and hence, to interpret data concerning the ME effect.

5.2.1 Uniaxial Structure

Let us apply these equations to the consideration of a ME material with a uniaxial structure having 3 m symmetry. In this case, 5.11 for the thermodynamic potential can be written as:

$$\begin{aligned}\Delta W_{ME} = {} & E_1[B_{11}(M_1^2 - M_2^2) - 2B_{22}M_1M_2 + 2B_{14}M_2M_3 \\ & + 2B_{15}M_1M_3] + E_1[B_{22}(M_2^2 - M_1^2) - 2B_{11}M_1M_2 + 2B_{15}M_2M_3 \\ & - 2B_{14}M_1M_3] + E_3(B_{33} - B_{31})M_3^2 + E_1^2[(b_{11} - b_{12})M_1^2 \\ & + (b_{13} - b_{12})M_3^2 + 2b_{14}M_2M_3 - 2b_{25}M_1M_3 + 2b_{16}M_1M_2] \\ & + E_2^2[(b_{11} - b_{12})M_2^2 + (b_{13} - b_{12})M_3^2 - 2b_{14}M_2M_3 + 2b_{25}M_1M_3 \\ & - 2b_{16}M_1M_2] + E_3^2(b_{33} - b_{31})M_3^2 + 2E_2E_3(2b_{41}M_1^2 + b_{41}M_3^2 \\ & + 2b_{44}M_2M_3 + 2B_{45}M_1M_3 + 2B_{52}M_1M_2) + 2E_1E_3(-2b_{52}M_1^2 \\ & - b_{52}M_3^2 - 2b_{45}M_2M_3 + 2b_{44}M_1M_3 + 2B_{41}M_1M_2) \\ & + 2E_1E_2(-2b_{16}M_1^2 - b_{16}M_3^2 + 2b_{25}M_2M_3 + 2b_{14}M_1M_3 + 2b_{66}M_1M_2);\end{aligned} \tag{5.17}$$

where the usual designations for two indexes is used: $B_{ijk} = B_{i\lambda}$ where $\lambda = 1, 2, 3, 4, 5, 6$ corresponds to $j = k = 1, j = k = 2, j = k = 3, j = 2$ and $k = 3, j = 1$ and $k = 3, j = 1$ and $k = 2$. Using this two indices system, we can then use either tensor $(3 \times 3 \times 3)$ or matrix (6×6) forms for the third rank linear ME coefficient.

Using effective demagnetization factors, the effective magnetic field can be calculated as (Bichurin et al. 2001)

$$\bar{H}_E = -\partial W_{ME}/\partial \bar{M} =_{-N} E_{\bar{M}}; \tag{5.18}$$

Equation 5.17 can be written in a system of coordinates $(1', 2', 3')$ for which the axis $3'$ coincides with the direction of spontaneous magnetization M_0, as shown in Fig. 5.2. The components of $H_{k'}^E$ can then be given as

$$H_{k'}^E = \beta_{k'kk}^E; \tag{5.19}$$

where the direction cosine matrix $\hat{\beta}$ follows from the system of coordinates in Fig. 5.2.

$$\hat{\beta} = \begin{pmatrix} 1 & 0 & 0 \\ 0 & \cos\Theta & -\sin\Theta \\ 0 & \sin\Theta & \cos\Theta \end{pmatrix} \tag{5.20}$$

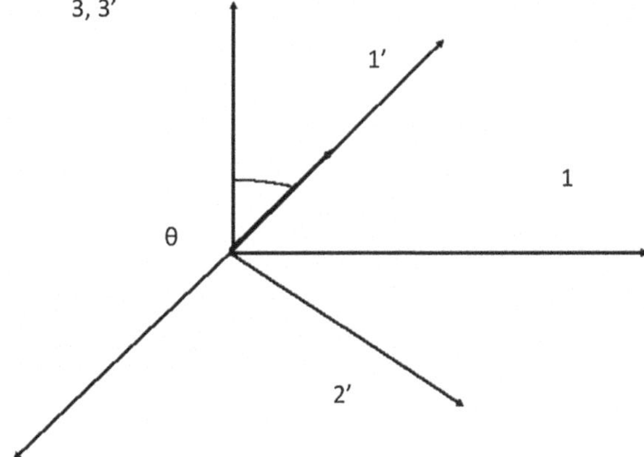

Fig. 5.2 System of coordinates for the multilayer composite

Accordingly, we also have that

$$M_K = \beta_{K'K} M_{K'}. \tag{5.21}$$

By substituting (5.18) into (5.19), and by considering (5.19)–(5.21), we can find that

$$\begin{aligned}
N_{11}^E - N_{33}^E &= 4(B_{11}E_1 - B_{22}E_2) + 2(b_{11} - b_{12})(E_1^2 - E_2^2) \\
&\quad + 8(b_{41}E_2E_3 - b_{16}E_1E_2 - b_{52}E_1E_3) + 2g_2\cos^2\Theta + 2g_3\sin2\Theta; \\
N_{22}^E - N_{33}^E &= 2g_2\cos2\Theta + 4g_3\sin2\Theta; \\
N_{12}^E &= [-2(B_{11}E_2 + B_{22}E_1) + 2b_{16}(E_1^2 - E_2^2) \\
&\quad + 4(b_{41}E_1E_3 + b_{52}E_2E_3 + b_{66}E_1E_2)]\cos\Theta \\
&\quad + [2(B_{14}E_2 - B_{15}E_1) + b_{25}(E_1^2 - E_2^2) - 4(b_{14}E_1E_2 \\
&\quad + b_{44}E_1E_3 + b_{45}E_2E_3)]\sin\Theta,
\end{aligned} \tag{5.22}$$

where

$$\begin{aligned}
g_2 &= -B_{11}E_1 + B_{22}E_2 + (B_{31} - B_{33})E_3 + (b_{12} - b_{13})E_1^2 \\
&\quad + (b_{11} - b_{13})E_2^2 + (b_{31} - b_{33})E_3^2 + 2(b_{16}E_1E_2 - b_{41}E_2E_3 \\
&\quad + b_{52}E_1E_3); \\
g_3 &= -B_{14}E_1 - b_{15}E_2 + b_{14}(E_2^2 - E_1^2) + 2(b_{25}E_1E_2 + b_{44}E_2E_3 \\
&\quad - b_{45}E_1E_3); \\
B_{66} &= (b_{11} - b_{12})/2.
\end{aligned}$$

Please note that indexes in the right-hand side of (5.22) correspond to those of the crystallographic system of coordinates. We then consider the case when the electric field is directed along the axis of symmetry: i.e., $E_1 = E_2 = 0$, and $E_3 = E$. For this specific orientation, we obtain

$$\begin{aligned} N_{11}^{E} - N_{33}^{E} &= 2\left[(B_{31} - B_{33})E + (b_{31} - b_{33})E^2\right]\cos^2\Theta; \\ N_{22}^{E} - N_{33}^{E} &= 2\left[(B_{31} - B_{33})E + (b_{31} - b_{33})E^2\right]\cos 2\Theta; \\ N_{12}^{E} &= 0. \end{aligned} \tag{5.23}$$

Taking into account the magnetic cubic anisotropy, and the anisotropy form for the ferrite layer oriented along the crystallographic (111) plane (i.e., $E \parallel H \parallel [111]$), it is possible to find that the resonance frequency depends on magnetic and electric fields, as given by

$$\omega/\gamma = H_3 + M_0[4/3 \times K_1/M_0^2 - 4\pi + 2(B_{31} - B_{33})E + 2(b_{31} - b_{33})E^2]; \tag{5.24}$$

where M_0 is now the effective saturation magnetization of the multilayer composite as an effective magnetic media. If we wish to only consider how the resonance frequency shifts under an applied electric field due to magnetoelectric interactions, we neglect all other contributions to (5.24) other than those provided via the linear ME coefficient B_{ijk}, which can be given as

$$\delta H_E = -2M_0(B_{33} - B_{31})E. \tag{5.25}$$

From 5.25 and 5.9, we can estimate the value of the ME coefficient in the microwave range (Bichurin et al. 2001): $2M_0(B_{31} - B_{33}) = 0.1$ A/kV for YIG–PZT composite, $2M_0(B_{31} - B_{33}) = 0.6$ A/kV for a lithium ferrite (LFO)–PZT composite, and $2M_0(B_{31} - B_{33}) = 1.4$ A/kV for a NFO–PZT composite. Similar analyses can be performed to determine the ME constants for other orientations of E and H fields.

Using the above values of the ME constant in the microwave range, it is now possible to define the changes in the magnetic susceptibility under an external electric field by 5.12. Results of such calculations of the magnetic susceptibility for bilayer composites are given in Figs. 6.3, 6.4, and 6.13. In Fig. 6.13, the dependence of the imaginary part of the magnetic susceptibility on dc bias is shown for a bilayer disc composite, where both constant magnetic and electric fields are perpendicular to the plane of the sample. For $E = 0$, a typical resonant line is observed. External constant electric field simply results in a shift of the resonant magnetic field, where the magnitude of the shift is proportional to the ME constants, which in turn are defined by the piezoelectric and magnetoelastic constants. Prior investigations have shown shifts of 330 Oe for NFO–PZT, and 22 Oe for YIG–PZT. The large ME effect in NFO is due to its much larger magnetostriction, in comparison to YIG. Clearly, the theoretical model presented above is an important tool for predicting ME interaction in the microwave frequency range. It allows us to define the ME constants, and their dependence on crystallography, material systems, and boundary parameters.

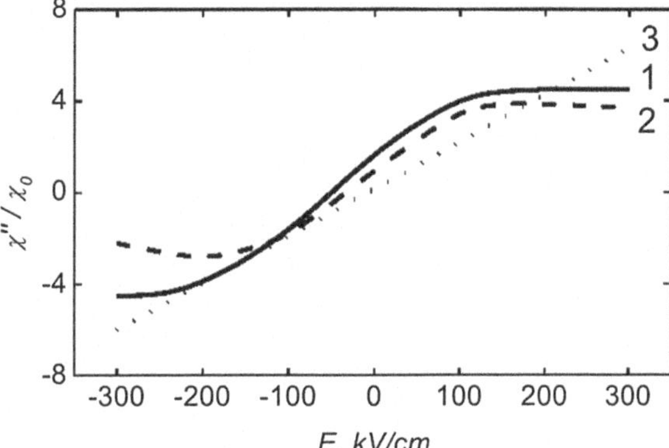

Fig. 5.3 Dependence of real part of magnetic susceptibility for layered structure at frequency 9.3 GHz on the electric field: *1* lithium ferrite–PZT, *2* nickel ferrite–PZT, *3* YIG–PZT

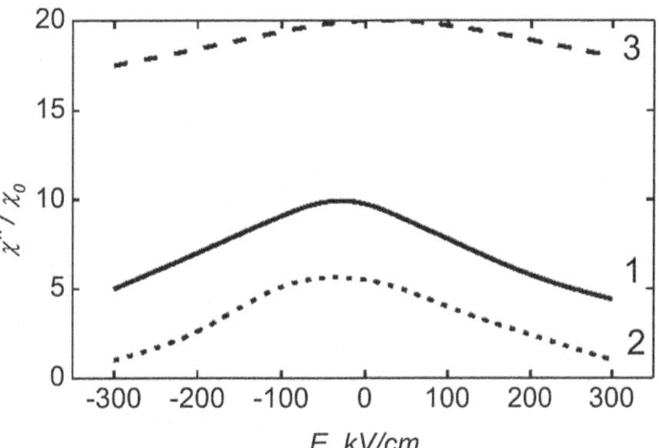

Fig. 5.4 Dependence imaginary part of magnetic susceptibility for layered structure at frequency 9.3 GHz on the electric field: *1* lithium ferrite–PZT, *2* nickel ferrite–PZT, *3* YIG–PZT

Figures 5.3 and 5.4 illustrate the dependence of the resonant (9.2 GHz) magnetic susceptibility on an external electric field. The width of the resonant line in terms of electric field is inversely proportional to the value of $2M_0$ (B_{31} − B_{31}). From 5.12, it then follows that a narrow resonant line is characteristic of a composite with strong ME effects. Accordingly, a bilayer composite of ferrite nickel–PZT has a narrower resonant line, relative to a similar composite structure with YIG–PZT.

The magnetic resonance data shown above taken under sweeping magnetic and electric fields is of possible interests for applications in solid state electronics. Electric field tunability of the magnetic subsystem allows for simplified methods for frequency tuning of microwave signals. In consideration that the dielectric breakdown strength of insulating thin layers that are well prepared can reach values of up to 300 kV/cm, it is possible to conceive of a ME resonator based on a simple bilayer composite structure of nicel ferrite–PZT that can be tuned by up to 925 MHz, and by 60 MHz for a similar structure of YIG–PZT.

5.3 Layered Composite with Single Crystal Components

ME interaction in ferromagnetic–ferroelectric heterostructures is mediated by elastic strain. Under an external electric field E, the piezoelectric component phase strains. This strain is then transferred to the ferrite component phase, resulting in a shift of its resonant magnetic field. Modeling of the microwave ME effect in layered or bulk composites has revealed two important types of ME interaction (Shastry et al. 2004; Bichurin et al. 2010), which are as follows: (i) between microwave magnetic and microwave electric fields and (ii) between microwave magnetic and constant electric fields.

There are practical difficulties in observing the bias dependence of the magnetic susceptibility shown earlier in Fig. 5.13. Measurements on samples of bulk composites of 90 %YIG–10 %PZT have shown only weak ME interactions, due to a low volume fraction of PZT; but, when the composite was loaded with higher volume fractions of PZT, a FMR line broadening masked the effect of an external electric field. Such line broadening, however, can easily be eliminated in layered structure by using single crystal films of YIG.

Next, let us discuss prior experimental results of microwave ME interactions in bilayer structures of YIG and lead magnesium niobate–lead titanate (PMN–PT) single crystals, and compare data to theoretical predictions. Single crystal heterostructures are preferable for experimental measurements because: (i) a smaller FMR line width facilitates exact definition of the resonant magnetic field shift, and accordingly the ME constants; and (ii) theoretically, the ME constant of single crystals should be larger relative to a corresponding polycrystalline sample (Bichurin et al. 1997). Bilayer structures (4×4 mm^2) were made using epitaxial YIG films of thickness 1–110 um grown on (111) gallium–gadolinium garnet (GGG) substrates that were 0.5 mm thick.

Measurements of the ME effect were then performed using an EPR spectrometer at a frequency 9.3 GHz. The input power was 0.1 mW, which corresponds to a magnetic field of about 130 A/м. The ME composite was placed outside of the resonator to eliminate potential overload of the resonator at the resonant absorption peak. Measurements were made under an electric field ($0 < E < 8$ kV/cm) applied perpendicular to the plane of composite for the following directions of magnetization: (i) H'', [011], and (ii) H'', (111).

For $E = 0$, the FMR line width was 0.5–4 Oe, depending on film thickness. Under external electric fields, the resonant magnetic field decreased with increasing E, as shown in Fig. 5.13. The dependence of resonant magnetic field shift on external dc electric field is shown in Fig. 5.5: data are given for H'', E'', for different thickness of YIG. A linear dependence of the resonant magnetic field shift on dc electric field can clearly be seen in this figure. The ME constant $A = dH_E/E$ can be estimated from data to be 5.4 A/kV for the composite with a YIG thickness of 4.9 um. Increase of the YIG thickness to 58 um resulted in a reduction of the ME constant to 4.4 A/kV.

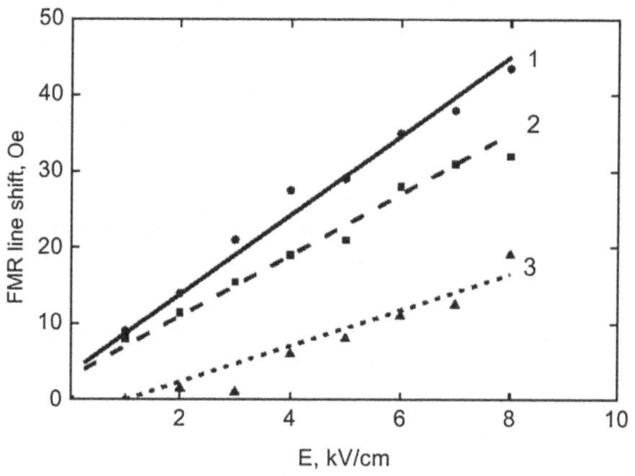

Fig. 5.5 Magnetoelectric effect in two-layer structures YIG (111) on GGG and PMN–PT (100) at frequency 9.3 GHz. Static fields E and H are parallel to an axis YIG [111] and perpendicular to sample plane. Shift of resonant magnetic field is shown as function E for different thickness of YIG film: 1—4.9 μm, 2—58 μm, 3—110 μm

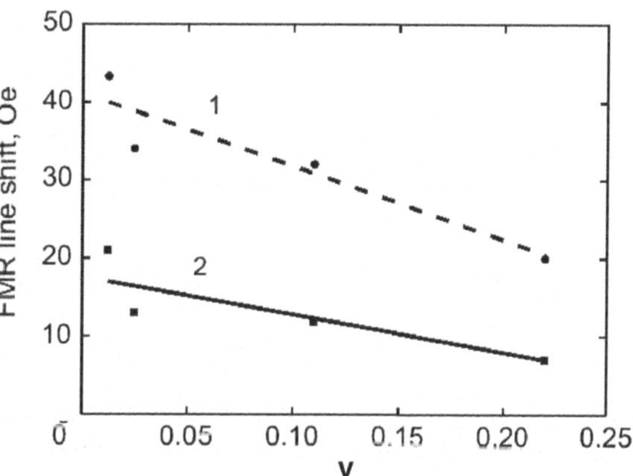

Fig. 5.6 Measured shift of resonant magnetic field in electric field $E = 8$ kV/sm as function of volume fractions of YIG and PMN–PT: 1—H'', [111], 2—H'', [011]

Measurements of the resonant magnetic field shift were also performed for H'', (011) plane of YIG. Along this direction, the resonant magnetic field shift is not as large as that when H'', (111). Accordingly, the values of dH_E and A were smaller for H'', (011). Prior investigations have been performed on magnetostrictive YIG and piezoelectric PMN–xPT bilayer composites as a function of the relative volume fractions. Figure 5.6 shows that the microwave ME effect depends near linearly on the relative volume fraction. The shift in the resonant magnetic field induced by an external electric field can be seen to decrease with increasing volume fraction of YIG, consistent with the predictions of the last section.

For comparisons of the experimental data with theoretical predictions, we should consider the additional term for the thermodynamic potential given by 5.11, which describes influence of external constant electric field E on the magnetic resonance spectrum. Let us then consider a bilayer composite of YIG and PMN–PT. First, we should calculate the deformation induced in PMN–PT by application of E. The generalized Hooke's law in this case, including piezoelectric constitutive relations, can be given as:

$$
\begin{aligned}
{}^{p}S_3 &= {}^{p}d_{33}E + {}^{p}s_{13}({}^{p}T_1 + {}^{p}T_2) + {}^{p}s_{33}\,{}^{p}T_3,\\
{}^{p}S_1 &= {}^{p}d_{31}E + {}^{p}s_{11}\,{}^{p}T_1 + {}^{p}s_{12}\,{}^{p}T_2 + {}^{p}s_{13}\,{}^{p}T_3,\\
{}^{p}S_2 &= {}^{p}d_{31}E + {}^{p}s_{12}\,{}^{p}T_1 + {}^{p}s_{11}\,{}^{p}T_2 + {}^{p}s_{13}\,{}^{p}T_3.
\end{aligned} \tag{5.26}
$$

Now, assume that the direction of the spontaneous polarization in the piezoelectric phase coincides with the [111] axis of the ferrite phase. In this case, the compliance tensor of the ferrite phase is

$$
{}^{m}s_{ijkl} = {}^{m}s_{i'j'k'l'}\beta_{ii'}\beta_{jj'}\beta_{kk'}\beta_{kk'}; \tag{5.27}
$$

where (1, 2, 3) is a system of coordinates in which the axis 3 is directed along direction of the spontaneous magnetization, and β is the direction of cosine matrix. For the ferrite phase we can then determine that

$$
\begin{aligned}
{}^{m}S_3 &= {}^{m}s_{13}({}^{m}T_1 + {}^{m}T_2) + {}^{m}s_{33}\,{}^{m}T_3,\\
{}^{m}S_1 &= {}^{m}s_{11}\,{}^{m}T_1 + {}^{m}s_{12}\,{}^{m}T_2 + {}^{m}s_{13}\,{}^{m}T_3,\\
{}^{m}S_2 &= {}^{m}s_{12}\,{}^{m}T_1 + {}^{m}s_{11}\,{}^{m}T_2 + {}^{m}s_{13}\,{}^{m}T_3.
\end{aligned} \tag{5.28}
$$

Next, let us use the usual boundary conditions for a mechanically free composite

$$
\begin{aligned}
{}^{p}S_1 &= {}^{m}S_1, {}^{p}S_2 = {}^{m}S_2, {}^{p}T_3 = {}^{m}T_3 = 0,\\
{}^{p}T_1 &= -{}^{m}v/{}^{p}v \times {}^{m}T_1, {}^{p}T_2 = -{}^{m}v/{}^{p}v \times mT_2;
\end{aligned} \tag{5.29}
$$

where ${}^{m}v$ and ${}^{p}v$ are the volume fractions of the ferrite and piezoelectric phases, respectively. Solutions of 5.26 and 5.28 by taking into account 5.29 gives for the stress on the magnetostrictive phase

$$
{}^{m}T_1 = {}^{m}T_2 = -\,{}^{p}d_{31}E\,{}^{p}v/[{}^{p}v({}^{m}s_{11} - {}^{m}s_{12}) + {}^{m}v({}^{p}s_{11} - {}^{p}s_{12})] \tag{5.30}
$$

The shift of the resonant magnetic field (which depends on the volume fractions of magnetic and piezoelectric phases) can then be calculated from 5.15 by taking into account 5.30 for the following two cases in the (111) plane of YIG: (i) H'', [111], and (ii) H'', [011].

Figure 5.7 illustrates the dependence of the ME constant A (which is numerically equal to the shift of the resonant line under $E = 1$ kV/cm) for a by-layer composite of YIG + GGG and PMN–PT. The thicknesses of the YIG + GGG and PMN–PT layers were 0.5 mm and 0.1 mm, respectively. Calculations were performed using the following material parameters: ${}^{p}d_{31} = -600 \times 10^{-12}$ m/V, ${}^{p}d_{33} = 1500 \times 10^{-12}$ M/V, ${}^{p}s_{11} = 23 \times 10^{-12}$ m^2/N, ${}^{p}s_{12} = -8.3 \times 10^{-12}$ m^2/N, $\lambda_{100} = -1.4 \times 10^{-6}$, $\lambda_{111} = -2.4 \times 10^{-6}$, $4\pi M_0 = 1750$ Gs, $H_a = -42$ Oe, ${}^{m}s_{11} = 4.8 \times 10^{-12}$ m^2/N, and ${}^{m}s_{12} = -1.4 \times 10^{-12}$ m^2/N. Figure 6.7 shows (i) that the ME constant decreases with increasing YIG volume fraction; and (ii) that

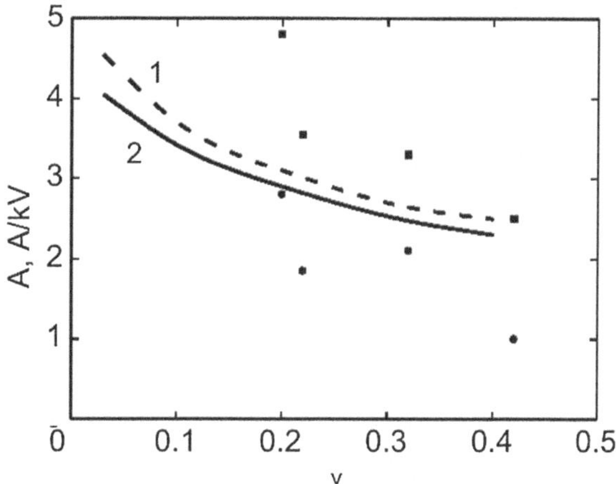

Fig. 5.7 Comparison of calculation (*line*) and data of ME constant for magnetic field laying in sample plane (*circles*) and perpendicular to sample plane (*squares*). Values of ME constant are presented as function of (YIG + GGG) and PMN–PT volume fractions

the ME interaction is larger for magnetic fields perpendicular to composite plane than for those parallel to it. In this figure, good qualitative and quantitative agreement can be seen between theory and data.

From the above results, we can conclude to obtain an optimum ME effect: (i) that the volume fraction of the piezoelectric phase should be high; (ii) that it is necessary to use a piezoelectric component phase with a large piezoelectric coefficient; and (iii) that it is necessary to use a magnetostrictive component phase with a small saturation magnetization and a high magnetostriction.

5.4 Resonance Line Shift by Electric Signal with Electromechanical Resonance Frequency

As we discussed in the last chapter, the electromechanical resonance (EMR) frequency range has a significant enhancement of the ME interaction. Here, we estimate the influence of an external electric field at the EMR frequency on the magnetic resonance spectrum for layered composite with a magnetostrictive ferrite component phase.

To determine the magnetic resonance line shift, we use (6.16). The calculation of the mechanical stress on the ferrite phase imposed by the piezoelectric on driven in the EMR range can be obtained from the equation of motion in 4.1, which has a general solutions of form given in (4.45) and (3.48). The integration constants can be obtained by assuming that the composite is mechanically free. In this case, the stress in the ferrite component phase can be given as

$$ {}^{m}T_1 = \frac{[\sin(kx)tg(kL/2) + \cos(kx)]E_3{}^{p}d_{31}v}{{}^{p}s_{11}(1-v) + {}^{m}s_{11}v}. \tag{5.31}$$

Srinivasan G, Tatarenko AS, Mathe V, Bichurin MI (2009) Microwave and MM-wave magnetoelectric interactions in ferrite-ferroelectric bilayers. Eur Phys J B 71:371–375

Tatarenko S, Bichurin MI (2012) Microwave magnetoelectric devices. Adv Condens Matter Phys 2012:286562

Tatarenko S, Srinivasan G, Bichurin MI (2006) Magnetoelectric microwave phase shifter. Appl Phys Lett 88:183507

Tatarenko AS, Bichurin MI, Gheevarughese V et. al (2010).Microwave magnetoelectric effects in ferrite-piezoelectric composites and dual electric and magnetic field tunable filters. J. Electroceram. 24:5

Zhai J, Li J, Viehland D, Bichurin MI (2007) Large Magnetoelectric susceptibility: the fundamental property of piezoelectric and magnetostrictive laminated composites. J Appl Phys 101:014102

Chapter 6
ME Effect at Magnetoacoustic Resonance Range

Abstract Resonant dependence of the ME voltage coefficient is investigated at overlapping the electromechanical and ferromagnetic resonances. A theoretical model predicts very strong ME interactions at magnetoacoustic resonance (MAR) in single-crystal ferrite-piezoelectric bilayer. In such bilayers, the ME interactions are mediated by mechanical strain. The theory predicts efficient transfer of energy between phonons, spin waves, and electric and magnetic fields at MAR. Ultrahigh ME coefficients, on the order of 80–480 V/cm Oe at 5–10 GHz, are expected for nickel ferrite-PZT and yttrium-iron garnet-PZT bilayers. Effects of exchange interactions on magnetoacoustic resonance are taken into account. We consider both direct ME effects and electric field induced magnetic excitations. The phenomenon is also of importance for the realization of multifunctional ME nanosensors/transducers operating at microwave frequencies.

Now, let us consider the bilayer composite as in Fig. 6.1 with ferrite and piezoelectric single crystal layers. The ferrite layer is supposed to be in a saturated single domain state (Bichurin et al. 2005, 2009; Ryabkov et al. 2006, 2007; Petrov et al. 2007, 2009). Bias field is assumed to be applied perpendicular to the sample plane. This state has two important advantages. First, when domains are absent, acoustic losses are minimum. Second, the single-domain state under FMR provides the conditions necessary for achieving a large effective susceptibility.

6.1 Direct Magnetoelectric Effect

The free energy density of a single crystal ferrite phase can be given as

$$^{m}W = W_H + W_{an} + W_{ma} + W_{ac}; \tag{6.1}$$

where $W_H = -\mathbf{M}\cdot\mathbf{H}$ is Zeeman's energy, $W_{an} = K_1/M^4(M_1^2\,M_2^2 + M_2^2\,M_3^2 + M_3^2\,M_1^2)$ is cubic anisotropy energy, K_1 is a constant of the cubic anisotropy, $W_{ma}= B_1/M^2\,(M_1^2\,{}^mS_1 + M_2^2\,{}^mS_2 + M_3^2\,{}^mS_3) + B_2/M^2(M_1\,M_2\,{}^mS_6 + M_2\,M_3\,{}^mS_4 + M_1\,M_3$

M. Bichurin and V. Petrov, *Modeling of Magnetoelectric Effects in Composites*, Springer Series in Materials Science 201, DOI: 10.1007/978-94-017-9156-4_6,

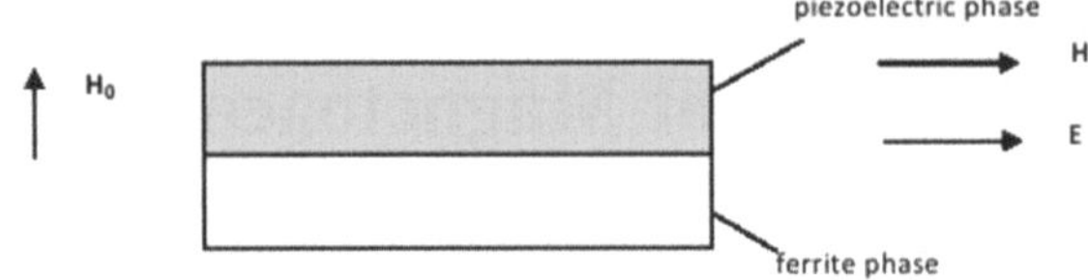

Fig. 6.1 Two-layer structure on the basis of single crystal phases

$^{m}S_5$) is the magnetoelastic energy, B_1 and B_2 are magnetoelastic constants, and $W_{ac} = ½\ ^{m}c_{11}(^{m}S_1^2 + {}^{m}S_2^2 + {}^{m}S_3^2) + ½\ ^{m}c_{44}\ (^{m}S_4^2 + {}^{m}S_5^2 + {}^{m}S_6^2) + {}^{m}c_{12}\ (^{m}S_1\ {}^{m}S_2 + {}^{m}S_2\ {}^{m}S_3 + {}^{m}S_1\ {}^{m}S_3$. In (6.1), it is supposed that the material is uniformly magnetized. On the basis of a generalized Hooke's law, we can write the stresses of the piezoelectric component phase as

$$\begin{aligned} {}^{p}T_4 &= {}^{p}c_{44}\,{}^{p}S_4 - e_{p15}\,{}^{p}E_2, \\ {}^{p}T_5 &= {}^{p}c_{44}\,{}^{p}S_5 - e_{p15}\,{}^{p}E_1; \end{aligned} \tag{6.2}$$

where e_{p15} is a shear piezoelectric coefficient. The equations of motion for the ferrite and piezoelectric phases are then

$$\begin{aligned} \partial^2(^{m}u_1)/\partial t^2 &= \partial^2(^{m}W)/(\partial x\partial\,^{m}S_1) + \partial^2(^{m}W)/(\partial y\partial\,^{m}S_6) \\ &\quad + \partial^2(^{m}W)/(\partial z\partial\,^{m}S_5), \\ \partial^2(^{p}u_1)/\partial t^2 &= \partial(^{m}T_1)/\partial x + \partial(^{m}T_1)/\partial y + \partial(^{m}T_1)/\partial z, \\ \partial^2(^{p}u_2)/\partial t^2 &= \partial(^{m}T_2)/\partial x + \partial(^{m}T_2)/\partial y + \partial(^{m}T_2)/\partial z. \end{aligned} \tag{6.3}$$

The equation of motion of the magnetization vector is given by

$$\partial \mathbf{M}/\partial t = -\gamma[\mathbf{M}, \mathbf{H_{eff}}], \tag{6.4}$$

where $\mathbf{H_{eff}} = -\partial(^{m}W)/\partial\mathbf{M}$.

To simply calculations, we assume waves with the circular polarization

$$\begin{aligned} m^+ &= m_1 + im_2, \\ H^+ &= H_1 + iH_2, \\ E^+ &= E_1 + iE_2, \\ u^+ &= u_1 + iu_2, \end{aligned} \tag{6.5}$$

where m is a variable magnetization, and u is the displacement. By taking into account (6.5), (6.3) and (6.4) take the form of

$$\begin{aligned} \omega m^+ &= \gamma(H_0m^+ - 4\pi\, M_0m^+ - M_0H^+), \\ -\omega^{2\,m}\rho^{m}u^+ &= {}^{m}c_{44}\partial^2(^{m}u^+)/\partial z^2, \\ -\omega^{2\,p}\rho\,^{p}u^+ &= {}^{p}c_{44}\partial^2(^{p}u^+)/\partial z^2. \end{aligned} \tag{6.6}$$

The boundary conditions at the interface between layers and at the top/bottom planes of the composite are given by

$$\begin{aligned} {}^{m}u^{+} &= {}^{p}u^{+} \text{ at } z = 0; \\ {}^{m}c_{44}\partial({}^{m}u^{+})/\partial z + B_2 m^{+}/M_0 &= 0 \text{ at } z = {}^{m}L; \\ {}^{m}c_{44}\partial({}^{m}u^{+})/\partial z + B_2 m^{+}/M_0 &= {}^{p}c_{44}\partial({}^{p}u^{+})/\partial z - {}^{p}e^{p}_{15}E^{+} \text{ at } z = 0; \\ {}^{p}c_{44}\partial({}^{p}u^{+})/\partial z - {}^{p}e^{p}_{15}E^{+} &= 0 \text{ at } z = -{}^{p}L; \end{aligned} \tag{6.7}$$

where ${}^{m}L$ and ${}^{p}L$ are the thicknesses of the ferrite and piezoelectric layers, respectively. Finally, the electric field E induced in the piezoelectric component can be defined by the open electric circuit condition

$$D^{+} = \frac{1}{{}^{p}L}\int_{-{}^{p}L}^{0} {}^{p}D^{+}dz = 0 \tag{6.8}$$

where ${}^{p}D^{+} = {}^{p}e_{15}{}^{p}S^{+} + {}^{p}\varepsilon_{11}{}^{p}E^{+}$ is the dielectric displacement of the piezoelectric layer.

Substitution of the solution of (6.6) into (6.8), and by taking into account (6.7), we can derive an expression for the ME voltage coefficient

$$\begin{aligned} |E^{+}/H^{+}| = {} & \gamma B_2\, {}^{p}c_{44}k_p\, {}^{p}e_{15}[1 - \cos(k^{p}_{p}L)]^{2}/\{(\omega - \gamma H_0 + 4\pi\gamma M_0) \\ & \times [-{}^{1}\!/\!_{2}\, {}^{p}c_{44}\, {}^{p}\varepsilon_{33}k_p\, {}^{p}L\sin(2k^{p}_{p}L)\,({}^{p}c_{44}k_p + {}^{m}c_{44}k_m) + (\cos(k^{p}_{p}L) - 1) \\ & \times {}^{p}e^{2}_{15}\left[\left(\cos\left(k^{p}_{p}L\right)({}^{m}c_{44}k_m + 2\,{}^{p}c_{44}k_p\right) + {}^{m}c_{44}k_m\right]\right]\}, \end{aligned} \tag{6.9}$$

$$\text{where } k_m = \omega\sqrt{\frac{{}^{m}\rho}{{}^{m}c_{44}}}, \quad k_p = \omega\sqrt{\frac{{}^{p}\rho}{{}^{p}c_{44}}}.$$

$$k_m = \omega\sqrt{\frac{{}^{m}\rho}{{}^{m}c_{44}}}, \quad k_p = \omega\sqrt{\frac{{}^{p}\rho}{{}^{p}c_{44}}}.$$

As follows from (6.9), there is a connection in the ferrite phase between the displacement and a homogeneous magnetization precession, through boundary conditions on the plate surfaces. Expression (6.9) shows that if the frequency of an applied magnetic field equals that of the magnetization precession (($\omega_0 = \gamma H_0 - 4\pi\gamma M_0$)), then the value of the ME voltage coefficient will be increased significantly. This enhancement is due to a coupling between the strain induced by an applied magnetic field in the range of the magnetic resonance and a corresponding one in the piezoelectric phase induced by an applied electric field.

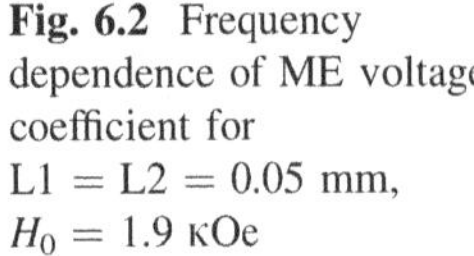

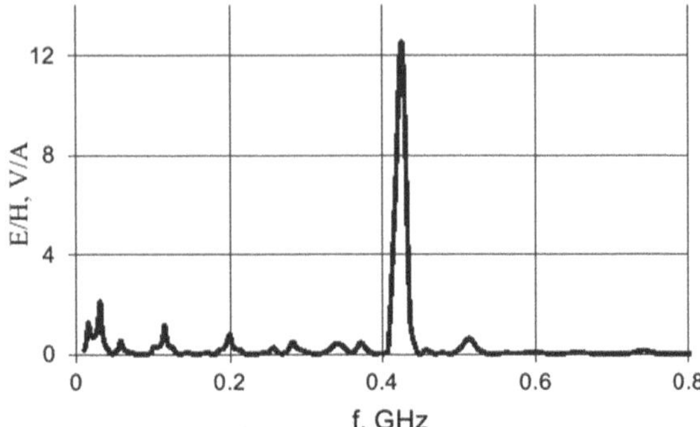

Fig. 6.2 Frequency dependence of ME voltage coefficient for L1 = L2 = 0.05 mm, $H_0 = 1.9$ кOe

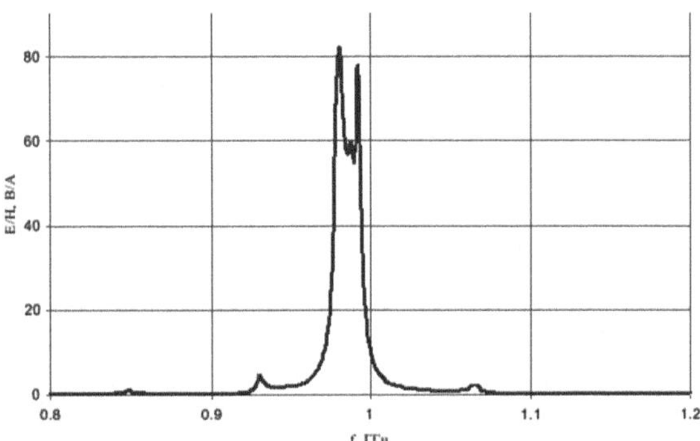

Fig. 6.3 Frequency dependence of ME voltage coefficient for L1 = L2 = 0.01 mm, $H_0 = 2.1$ кOe

Figures 6.2 and 6.3 show the dependences of the ME voltage coefficient for a bilayer composite of single crystal YIG and PMN–PT, both of which were calculated by (5.9). Calculations were performed by introducing a complex frequency to account for magnetoacoustic loss. The real component of the frequency was taken as $\omega_r = 10^7$ rad/s.

For frequencies less than that of the homogeneous magnetization precession, the microwave ME voltage coefficient has a maximum when driven by an electric field in the EMR frequency range. Essential to this increase in the microwave ME voltage coefficient is that the EMR frequency of the electric be equal to that for a uniform magnetization precession in the ferrite phase. Even after including magnetoacoustic loss factors (as mentioned above), the microwave ME coefficient can reach giant values of up to 64 V/(cm Oe).

6.2 Effects of Exchange Interactions on Magnetoacoustic Resonance

It is well known that the influence of exchange interactions on magnetoelastic waves is stronger with frequency. The exchange interaction determines the strength of the spin wave stiffness parameter and the frequency of the magnetoelastic modes. This section is concerned with the effect of magnetic exchange interactions on ME coupling at MAR in a ferrite-piezoelectric nanobilayer (Bichurin et al. 2005).

The piezoelectric phase of thickness L_1 is supposed to be poled with an electric field E_0 along x direction and the in-plane microwave electric field E is along z. The bias magnetic field H_0 is along z and the alternating magnetic field h (along $-y$) is assumed to be tangential to the bilayer plane. We first solve the equations of motion for magnetization for the ferrite and equations of mechanical displacements for the piezoelectric and ferrite.

The equations of media motion for the ferrite and piezoelectric phases and equation of motion of magnetization take the form taking into account exchange interactions:

$$\begin{aligned} i\omega\,\tilde{m}_x &= -\gamma\left(H_0\tilde{m}_y - H_a a^2\frac{\partial^2\tilde{m}_y}{\partial x^2}\right),\\ i\omega\,\tilde{m}_y &= \gamma\left((H_0+4\pi M_s)\tilde{m}_x - H_a a^2\frac{\partial^2\tilde{m}_x}{\partial x^2} + B_2\frac{\partial U_{mz}}{\partial x}\right),\\ -\rho_m\omega^2 U_{mz} &= c_{m44}\frac{\partial^2 U_{mz}}{\partial x^2} + \frac{B_2}{M_0}\frac{\partial\tilde{m}_x}{\partial x},\\ D_z &= \varepsilon_p E_z + e_{p15}\frac{\partial U_{pz}}{\partial x},\\ -\rho_p\omega^2 U_{pz} &= c_{p44}\frac{\partial^2 U_{pz}}{\partial x^2}. \end{aligned} \tag{6.10}$$

where H_a is exchange field, and a is lattice constant for YIG. Next we let $M_z = M_s$ and $h_x = 0$ since the ac field magnetic field is along y-axis. Besides, for given composite orientation we have $N_x = 4\pi$, $N_y = N_z = 0$.

Solving equations of motion of magnetization and media motion enables finding displacement of layers and induced voltage in piezoelectric layer at open circuit condition. For ME voltage coefficient, one obtains the following expression:

$$\begin{aligned} \alpha_E =\,& 4\pi i\,\omega\,\gamma\,e_{p15}\,c_{p44}k_p\,k_m B_2\\ &\times\left(1-\cos k_p L_1\right)\left(\cos k_m L_2 - 1\right)/\left[\varepsilon_{p33}c_{p44}k_p L_1\left(\rho_m\omega^2\sin k_m L_2\cos k_p L_1\right.\right.\\ &\left.+\,c_{p44}k_p\,k_m\,\sin k_m L_2\cos k_p L_2\right) + 4\pi e_{p15}^2\left(\rho_m\omega^2\sin k_p L_1\cos k_m L_2\right.\\ &\left.\left.+\,2c_{p44}k_p k_m\left(1-\cos k_p L_1\right)\cos k_m L_2\right)\left(\omega^2-\gamma^2(H_0+4\pi M_s)\,H_0\right)\right]. \end{aligned} \tag{6.11}$$

where L_1 and L_2 are thicknesses of piezoelectric and ferrite phases, $k_m = \omega\sqrt{\rho_m}\left[c_{m44} + \frac{\gamma^2[B_2^2 H_0 + 2M_s a^2 H_a\omega^2\rho(H_0+2M_s)]}{M_s[\omega^2-\gamma^2 H_0(H_0+4M_s)]}\right]^{-1/2}$, $k_p = \omega\sqrt{\frac{\rho_p}{c_{p44}}}$.

The theory developed here is applied to a YIG-PZT bilayer. The thicknesses of PZT and YIG layers are assumed to be 100 nm and 195 nm, respectively, so that the EMR thickness mode in PZT will have the same frequency as for FMR in YIG.

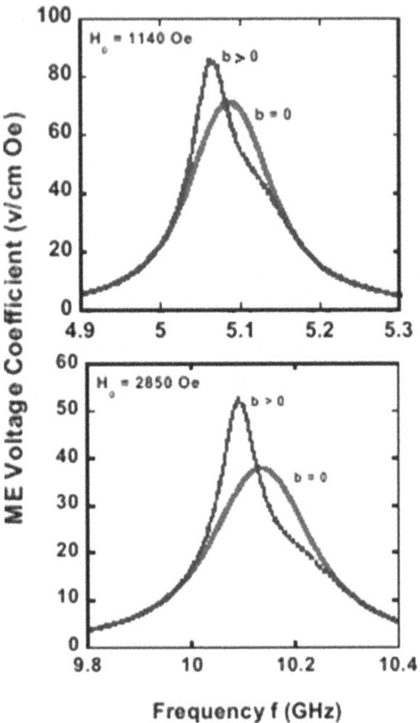

Fig. 6.4 The ME voltage coefficient α_E versus frequency profile for a bilayer of lead zirconate titanate (PZT) of thickness 100 nm and yttrium iron garnet (YIG) of thickness 195 nm. Estimates are for magnetic exchange b = 0 and b ≠ 0 and dc magnetic fields of 1140 and 2850 Oe

Figure 6.4 shows the variation in α_E with f. The results are for bias fields $H_0 = 1140$ Oe for coincidence of FMR and fundamental EMR, and $H_0 = 2850$ Oe for the coincidence of FMR and higher harmonics of EMR. Consider the results for a bias field of 1140 Oe. A symmetric profile with a peak α_E of 65 V/cmOe is seen when the exchange parameter $H_a = 0$. When the effect of exchange is considered, one notices a sharpening of the resonance profile with a 30 % increase in ME coeffcient and a down shift in the resonance frequency. One obtains the "exchange amplification" of ME effect. The appearance of a high frequency shoulder is indicative of flow of rf energy to magnetoacoustic modes. Similar effects are expected for a frequency of about 10 GHz when the bias field is increased to 2850 Oe. When exchange interaction is considered, Fig. 6.4 reveals a 60 % increase in peak α_E and a broad shoulder due to magnetoacoustic excitations.

Figure 6.5 shows the effect of PZT-to-YIG thickness on ME coupling. The results also correspond to volume dependence of α_E since the films are assumed to have equal surface areas. For a thickness ratio of 0.74, the figure shows a shallow and broad resonance. When the PZT thickness is reduced so that the thickness ration is 0.32, one notices a sharp magnetoacoustic resonance and the shoulder due to spin wave modes is clearly resolved. With a further decrease in PZT thickness, there is reduction in peak α_E and an up shift in the spin-acoustic wave frequencies.

Results on thickness dependence of peak—$\alpha_{E\ E}$ obtained from profiles as in Fig. 6.5 are shown in Fig. 6.6. For small PZT thickness, $\alpha_{E\ E}$ remains small and shows an increase with increasing PZT thickness. Since EMR in PZT facilitates

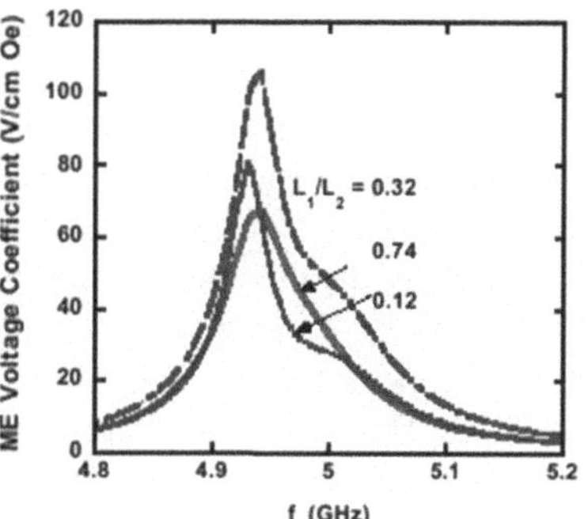

Fig. 6.5 Results as in Fig. 6.4, but as a function of ratio of PZT (L_1) to YIG (L_2) thickness. The static magnetic field is 1100 Oe. The total thickness is half-wave length

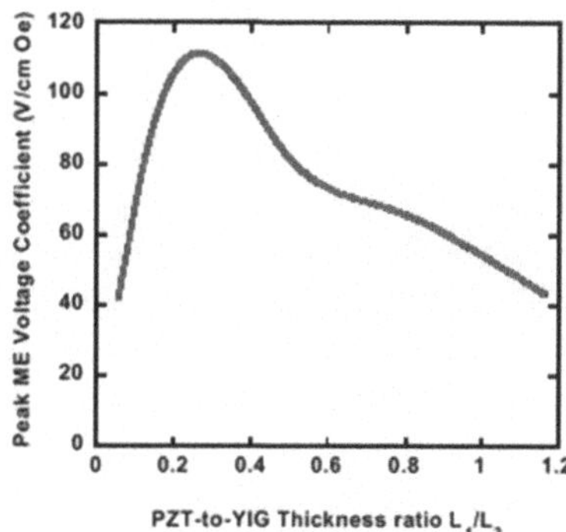

Fig. 6.6 Variation of peak ME voltage coefficient α_E with the PZT to YIG thickness ratio. The bias field H0 = 1100 Oe, L1 + L2 = half-wave length

the maximum transfer of rf power to magnetic subsystem, increase in α_E with PZT thickness is expected. But for thickness ratios above 0.32, we see a drop in the ME coefficient. Thus maximum ME coupling is expected when the YIG layer is three times as thick as the PZT.

The theory discussed here and estimates for YIG-PZT bilayers will be a roadmap for experiments on magnetoacoustic resonance and investigations on the effects of magnetic exchange interactions. Such bilayers and ME phenomenon are potentially useful for devices such as resonators or phase shifters based on magnetoacoustic waves.

6.3 Electric Field Induced Magnetic Excitations

This section is concerned with consideration of magnetic excitations in a ferrite-piezoelectric bilayer due to microwave electric field and ME interactions (Petrov et al. 2007; Bichurin et al. 2010). The magnetic response is described in terms of ME susceptibility and a novel technique has been proposed for its determination for a yttrium iron garnet (YIG)-lead zirconate titanate (PZT) nanobilayer. It is assumed that the sample is positioned at the maximum of microwave electric field.

An induced microwave magnetic field (parallel to the electric field) will result from ME interactions and, therefore, will lead to magnetic excitations in the bilayer. Such magnetic excitations originate from elastic modes in the piezoelectric component. These acoustic modes would in turn excite coupled magnetoelastic modes in the ferrite due to ME coupling. The excitations are standing waves along the thickness of the sample and the wave length is determined by the thickness of PZT and YIG, and materials parameters. These coupled magnon-phonon modes will be in the microwave region of the electromagnetic spectrum for YIG. Thus the focus here is high frequency magnetic excitations, including ferromagnetic resonance (FMR) and ME susceptibility in a ferrite-ferroelectric bilayer. Traditional FMR at high powers in a ferrite will lead to nonlinear effects such as saturation of main resonance and subsidiary absorption. The idea here is to eliminate those effects by locating a bilayer at the position of maximum rf electric field.

We consider a ferrite-PZT bilayer as in Fig. 10 that is subjected to a bias field $\boldsymbol{H_0}$ perpendicular its plane, along the z-axis. The piezoelectric phase is electrically polarized with a field $\mathbf{E_0}$ parallel to z. The expression for the space-variant microwave magnetization m_k can be obtained by solving combined equations of medium motion for ferrite and piezoelectric phases and equation of motion of magnetization for the ferrite. Thus $\partial m/\partial z$ will appear in the equation for mechanical displacement for ferrite, and the term $\partial u/\partial z$ will be present in the equation of motion of magnetization. The magnetization m_k in terms of circularly polarized mechanical displacement ${}^{m}u^{+}$ of the ferrite is obtained by solving the above equations and substituting in

$$m_k = \frac{B_2 \gamma \frac{\partial({}^{m}u^{+})}{\partial z}}{\omega - \omega_k} \tag{6.12}$$

The magnetic modes will have uniform magnetization in the plane of the film and a standing wave structure perpendicular to the film plane. Substituting the value for ${}^{m}u^{+}$ into Eq. (60) yields:

$$m_k = \frac{B_2 \gamma e_{15}{}^{p}E^{m}k \, \sin[{}^{m}k({}^{m}L - z)\,]\,[1 - \cos({}^{p}k\,{}^{p}L)]}{[{}^{p}c_{44}{}^{p}k \, \sin({}^{p}k\,{}^{p}L)\cos({}^{m}k{}^{m}L) + {}^{m}c_{44}{}^{m}k \, \sin({}^{m}k\,{}^{m}L)\cos({}^{p}k{}^{p}L)](\omega - \omega_k)} \tag{6.13}$$

where ${}^{m}c_{44}^{+} = {}^{m}c_{44} + \gamma(B_2^2 + \omega^2\,{}^{m}\rho M_0 H_e)/[M_0(\omega - \gamma H_0 - 4\pi\gamma M_0)$,

$${}^{m}k = \omega\sqrt{\frac{{}^{m}\rho}{{}^{m}c_{44}^{+}}}, \quad {}^{p}k = \omega\sqrt{\frac{{}^{p}\rho}{{}^{p}c_{44}}}.$$

Signal attenuation is taken into account by introducing a complex frequency and an imaginary component of $\omega'' = 10^{-3}\,\omega_k$. This imaginary component corresponds to a Q-value of 1000 for resonance absorption in the ferrite.

Next we apply the theory to the specific case of YIG-PZT bilayer and calculate the ME susceptibility given by $\alpha = \mu_0 \,\partial m_k/\partial\, {}^{P}E$. The choice of YIG for the ferrite is because of low-losses at microwave frequency, a necessary condition for the observation of the enhancement in the ME coupling that is predicted by the theory. Figure 17a shows the susceptibility versus frequency f. We choose 100 nm YIG and 195 nm PZT so that the fundamental electromechanical resonance (EMR) will be around 6 GHz. A bias filed of $H_0 = 2$ kOe is assumed so that it is smaller than the field H_r for the excitation of magnetic modes that include FMR. There are peaks in α at the fundamental EMR and higher order thickness modes. The susceptibility at the fundamental mode is an order of magnitude higher than the value at the higher harmonics.

Consider the results in Fig. 6.7b for a bias field H_0 corresponding to magnetic resonance in YIG. When H_0 is set equal to $H_r = \omega/\gamma + 4\pi M_{0,\alpha}$ is expected to show a dramatic increase in magnitude, as in Figs. 6.7b and c, due to coincidence of resonance character for the mechanical displacement and magnetization. When the frequencies of magnon and phonon modes are matched, there is efficient transfer of energy between the electric and magnetic subsystems. In Fig. 6.7b for $H_0 = 3.86$ kOe, the fundamental acoustic mode coincides with uniform precession magnon mode that results in a 60-fold increase in α. When the H_0 is increased to 6 kOe so that uniform precession frequency coincides with the higher order EMR mode, one expects two orders of magnitude increase in α as in Fig. 6.7c.

The z-dependence of the ME susceptibility is shown in Fig. 6.8. The results for 100 nm YIG—195 nm PZT bilayer shows the average value of the ME susceptibility at each z, with z = 0 representing the interface. As one moves along +z away from the interface, the susceptibility decreases linearly to zero on the outer surface of YIG.

Next we consider measurements of ME susceptibility for the case of a PZT-YIG bilayer. One could use a resonant cavity with the sample located at the ac electric field maximum. The dc magnetic bias field is selected so that homogeneous precession frequency coincides with the fundamental EMR mode. Thus the microwave electric field will result in FMR in YIG and absorption of microwave power. The electric field induced magnetization is equivalent to that induced by a microwave magnetic field

$$H = m_k \frac{\gamma H_0 - 4\pi\gamma M_0 - \omega}{\gamma M_0}. \tag{6.14}$$

Equation (6.14) is obtained by solving the equation of motion of magnetization for a YIG plate being placed in an antinode of ac magnetic field. The absorbed power P is given by $P = k_1 H^2$ with $k_1 = \frac{\pi M_0}{\Delta H}\omega V$, where V is volume of YIG and ΔH is half-width of resonance absorption. Thus, H can be determined from (6.14) and substituted in (6.13) to obtain m_k. The ME susceptibility can be determined from data on power absorbed.

For using ME composites in microwave devices, the electronic noise is a crucial issue. Noise can be produced by several different effects. Thermal noise and shot

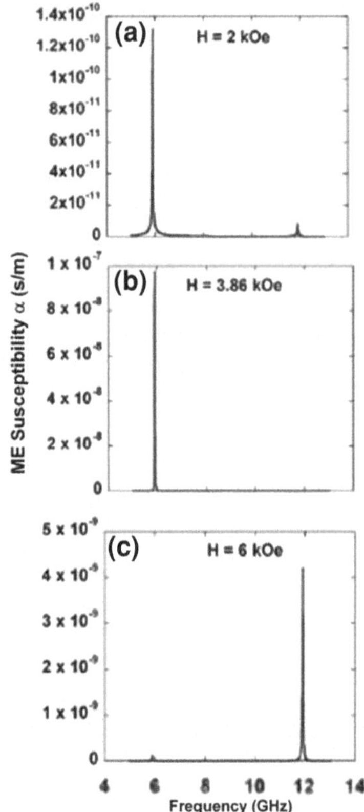

Fig. 6.7 Theoretical frequency dependence of the ME susceptibility at the ferrite-piezoelectric interface ($z = 0$) for a bilayer of YIG and PZT subjected to dc and ac fields. The bilayer is assumed at the microwave electric field maximum. **a** Results are for $H_0 = 2$ kOe that is well below the magnetic resonance field for YIG. The peaks correspond to elastic modes in PZT. **b** ME susceptibility for $H_0 = 3.86$ kOe, corresponding to coincidence of fundamental elastic mode in PZT and magnetic resonance in YIG. **c** Similar data at $H_0 = 6$ kOe, at coincidence of higher order elastic mode and magnetic modes

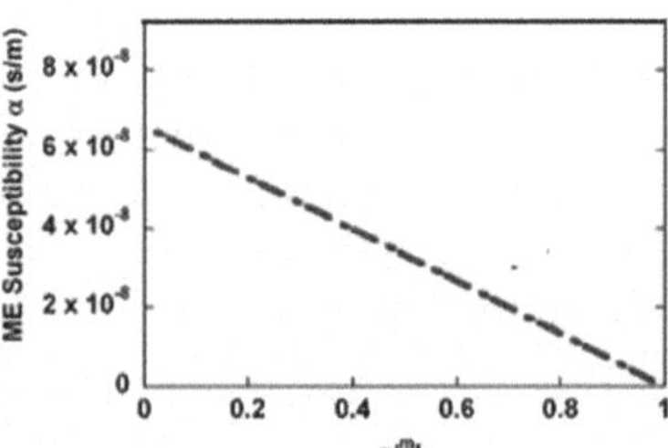

Fig. 6.8 The average ME susceptibility versus z for a YIG-PZT bilayer, with $z = 0$ is the interface and ^mL is the thickness of YIG. The susceptibility decreases from a maximum value to zero along the thickness of YIG

noise are known to be inherent to all electronic devices. Flicker noise, also known as 1/f noise, can be produced by ferrite-piezoelectric composites. Using the equivalent circuit of composite in the form of capacitor C connected in parallel with the loss resistance R which produces noises, the following expressions for C and R can be obtained: $C = \varepsilon' A/d$ and $R = 1/(\omega C \tan \delta)$, where ε', A, d and δ are real part of permittivity, cross-sectional area, sample thickness and dielectric loss angle, correspondingly. The dielectric loss angle was recently reported for the

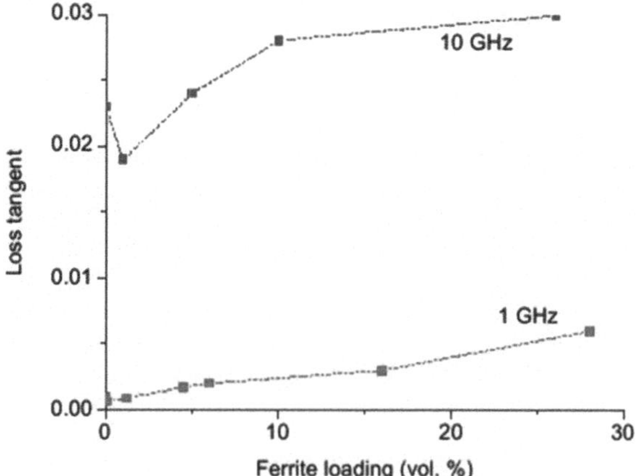

Fig. 6.9 The ferrite volume fraction dependence of loss angle tangent for frequency of 1 and 10 GHz and the composite of nickel-zinc spinel ferrite and barium strontium titanate

composite of nickel-zinc spinel ferrite and barium strontium titanate. The ferrite volume fraction dependence of loss angle tangent is shown in Fig. 19 for frequency of 1 and 10 GHz (Fig. 6.9).

6.4 Conclusion

A theoretical model has been discussed for ME effects in a single-crystal ferrite-piezoelectric bilayer in the magnetoelastic resonance region. The theory predicts giant ME effect at MAR. The enhancement arises from interaction between elastic modes and the uniform precession spin-wave mode, resulting in magnetoelastic modes. The peak ME voltage coefficient appears at the coincidence of acoustic resonance and FMR frequencies. Estimates for nominal bilayer parameters for nickel ferrite-PZT and YIG-PZT predict MAR at 5–10 GHz and ME voltage coefficient on the order of 80–480 V/(cm Oe) taking into account exchange interaction. In our calculations, we suppose that layer thickness is sufficiently large to neglect the influence of strain relaxation on average stresses in the structures that determine the ME voltage coefficient. This is valid for layers thickness of 30–50 nm or more. If necessary, the lattice mismatch effect can be taken into account by using the Landau–Ginsburg–Devonshire phenomenological thermodynamic theory.

Obtained estimates will certainly be of interest for potential device applications based on multilayer composites with ME-effects. Microwave devices based on ME effects have unique advantages over traditional ferrite and semiconductor analogues (Bichurin 1990, 1994). In general, the control can have micro- and macro-nature and be connected with the change of electric field strength, tension of mechanical field, magnetic field strength, temperature and their combined effect and also with change of medium activity conditions.

References

Bichurin MI (1990) Resonance microwave magnetoelectric effect: magnetoelectric substances. Nauka, Moscow, pp 53–67 (in Russian)

Bichurin MI (1994) Magnetoelectrics in microwave range. Ferroelectrics 161:53–57

Bichurin MI, Petrov VM, Ryabkov OV, Averkin SV, Srinivasan G (2005) Theory of magnetoelectric effects at magnetoacoustic resonance in single-crystal ferromagnetic-ferroelectric heterostructures. Phys Rev B 72:060408

Bichurin MI, Petrov VM, Averkin SV, Liverts E (2010) Present status of theoretical modeling the magnetoelectric effect in magnetostrictive-piezoelectric nanostructures. Part II: magnetic and magnetoacoustic resonance ranges. J Appl Phys 107:053905

Bichurin MI, Petrov VM, Averkin SV, Filippov AV, Liverts E, Mandal S, Srinivasan G (2009) Modelling of magnetoacoustic resonance in ferrite–piezoelectric bilayers. J Phys D: Appl Phys 42:215001

Petrov V, Srinivasan G, Ryabkov OV, Averkin SV, Bichurin MI (2007) Microwave magnetoelectric interactions in ferrite–piezoelectric nanobilayers: theory of electric field induced magnetic excitations. Solid State Commun 144:50–53

Petrov VM, Zibtsev VV, Srinivasan G (2009) Magnetoacoustic resonance in ferrite-ferroelectric nanopillars. Eur Phys J B 71:367–370

Ryabkov OV, Petrov VM, Bichurin MI, Srinivasan G (2006) Magnetoacoustic resonance in tangentially magnetized ferrite-piezoelectric bilayers. Tech Phys Lett 32:1021–1023

Ryabkov OV, Averkin SV, Bichurin MI, Petrov VM, Srinivasan G (2007) Effects of exchange interactions on magnetoacoustic resonance in layered nanocomposites of yttrium iron garnet and lead zirconate titanate. J Mater Res 22:2174–2178

Chapter 7
Conclusions

We discussed here the ME properties of ferrite–piezoelectric composites, to create new ME composites with enhanced ME couplings that would enable them for application in functional electronics devices. To address this important scientific and technical goal, a generalization of various theoretical and experimental studies of ME composites has been given. One of the main tasks according to the formulated approach is a comparative analysis of ME composites that have different connectivity types. The relative simplicity of manufacturing multilayer composites with a 2–2 type connectivity having giant ME responses is an important benefit. In addition, composites with 3–0 and 0–3 connectivity types can also be made in considerable quantity by a minimum monitoring of the synthesis process.

Availability of theoretical models for composite properties is necessary to interpret experimental data and to restrict oneself among the multitude of composite configurations. Theoretical estimations of the ME voltage coefficient for series and parallel composite models, and also a cubic model for composites with a 3–0 connectivity type, are known. However, as already mentioned, ME voltage coefficient was computed as the ratio of an electric field induced in the piezoelectric phase to the magnetic field applied to the magnetic one: i.e., $\alpha_E = {}^pE/{}^mH$. But, in reality, the internal fields in composite components can be significantly different from the external fields. In particular, formulas predicted a maximum ME voltage coefficient in a pure piezomagnetic phase (i.e., ${}^pv = 0$): which distinctly mismatches reality. In addition, generalized models for composite based on an effective medium method were presented. These offer determination of the effective composite parameters with phase connectivity types of 2–2, 3–0, and 0–3 that are based on exact solutions. The ultimate purpose of any theoretical work must be to predict the ME parameters—both susceptibility and voltage coefficients—as these are the most basic parameters of magnetoelectricity.

To describe the composite's physical properties, the exact solution of elastostatic and electrostatic equations were obtained. Expressions for the ME susceptibility and ME voltage coefficient were derived as functions of an interface coupling parameter, constituent phase material parameters, and relative volume fractions of phases. Longitudinal, transverse and in-plane cases were all

M. Bichurin and V. Petrov, *Modeling of Magnetoelectric Effects in Composites*, Springer Series in Materials Science 201, DOI: 10.1007/978-94-017-9156-4_7,

considered. For a bilayer that is an asymmetric structure, the influence of flexural deformations of sample on ME output was estimated.

It is shown that the ME coupling in the EMR region exceeds the low-frequency value by more than an order of magnitude. It was found that the peak transverse ME coefficient at EMR is larger than the longitudinal one. This is accounted for (i) by higher energy losses for longitudinal fields orientation due to eddy currents in the electrodes and (ii) by influence of demagnetization field that appears at longitudinal orientations that reduces the piezomagnetic modulus. Predictions of the ME effect for various model composite systems were given including CFO–PZT and lanthanum strontium manganite–PZT. It was shown that ME effect in ferrite–PZT systems is maximum for in-plane magnetic and electric fields. The theoretical estimates of ME parameters were compared with experimental data.

The specific focus is on ME coupling at magneto-acoustic resonance, i.e., at the coincidence of electromechanical resonance in the piezoelectric phase and ferromagnetic resonance in a tangentially magnetized ferrite. When exchange is ignored, estimated ME coefficient versus frequency profile shows a giant magnetoelectric coefficient at MAR. Exchange interactions will facilitate flow of microwave energy to magnetoacoustic spin waves, leading to the presence of a secondary peak in α_E versus f profiles. Estimates of the ME coefficient with and without exchange as and as a function of layer thickness are discussed for PZT–YIG bilayers.

ME effects occur over a broad frequency bandwidth, extending from the quasi-static to millimeter ranges. This offers important opportunities in potential device applications. It makes possible new concepts in sensing, gyrators, microwave communications, phase shifters, just to name a few. It also complicates the understanding of magnetoelectricity, as there are significant changes in its spectra with frequency. There are strong enhancements in the ME coefficients near both the electromechanical and magnetic resonances. In addition, there is the important problem of studying dispersion in the ME parameters over a broad frequency range of $10^{-3} < f < 10^{10}$ Hz. Relaxation parameters depend on connectivity type, composite geometry and structure, and volume fraction of constituent phases.

Index

M. Bichurin and V. Petrov, *Modeling of Magnetoelectric Effects in Composites*, Springer Series in Materials Science 201, DOI: 10.1007/978-94-017-9156-4,

The manufacturer's authorised representative in the EU is Springer Nature Customer Service Centre GmbH, Europaplatz 3, 69115 Heidelberg, Germany. If you have any concerns regarding our products, please contact ProductSafety@springernature.com

Printed and bound by CPI Group (UK) Ltd, Croydon, CR0 4YY

15/07/2026

02167644-0002